Protected Cultivation of Flowers

NIPA® GENX ELECTRONIC RESOURCES & SOLUTIONS P. LTD.
New Delhi-110 034

Protected Cultivation of Flowers

Sachin Tyagi
Sanjay Sahay
Department of Horticulture (F & FT)
Bihar Agriculture University
Sabour, Bhagalpur – 813210 (Bihar) India

NIPA® GENX ELECTRONIC RESOURCES & SOLUTIONS P. LTD.
New Delhi-110 034

NIPA® GENX ELECTRONIC RESOURCES & SOLUTIONS P. LTD.

101,103, Vikas Surya Plaza, CU Block
L.S.C. Market, Pitam Pura, New Delhi-110 034
Ph : +91 11 27341616, 27341717, 27341718
E-mail: newindiapublishingagency@gmail.com
web: www.nipabooks.com

For customer assistance, please contact
Phone: + 91-11-27 34 17 17
Fax: + 91-11- 27 34 16 16
E-Mail: feedbacks@nipabooks.com

ISBN: 978-81-19254-45-3

Composed and Designed by NIPA®.

Preface

Off late floriculture has been gaining a lot of attraction btoh in terms production and sales from the scientists and farmers point of views.

This has forced the grower to move and adopt more of technological option to increase the production.

Adopting protected cultivation techniques is one such means and the book caters to this, with information contributed by known scientists and faculty members across the country.

Specific crops have also been covered and explained with ease. Easy language will help students to understand the information given.

I would like to thank the Vice Chancellor of BAU for his constant encouragement.

Editors

Contents

Contents

1

History and Overview of Protected Cultivation

N. Basavaraja, Suhasini Chikkalaki, Viresh Hiremath, Mallikarjun Awati and Miss Suhasini

Office of the Dean Post Graduate Studies, University of Horticultural Sciences Bagalkot, Karnataka-587104

Introduction

'Protected cultivation' is a method of cultivation of plants or crops in artificial environment (Polyhouse/ Greenhouse/Glasshouse) to protect the plants or crops from the natural calamities *viz.* wind, hailstorm, excessive radiation, temperature extremes, insect-pests and diseases. For matching ever growing needs of human population in the country, it has become imperative to provide the not just cereals but also fruits, vegetables, flowers and other horticultural products to fulfill their requirements. In the present scenario of perpetual demand of horticultural produce in order to increase their production round the year and shrinking landholdings drastically, protected cultivation is the best alternative and drugery- less approach for using land and other resources more effeciently.Protected cultivation comprised of selection of crop and media, climate control,marketing, post harvest technology, plant protection and irrigation/ fertigation.

In India, states like Karnataka, Andra Pradesh, Maharashtra and Tamil Nadu are the best suited for protected cultivation of horticultural crops due to their pleasent climatic conditions. Apart from these states, Himachal Pradesh, Uttar Pradesh, Jammu and Kashmir, West Bengal, Delhi and Jharkhand are also fore runners in adopting protected cultivation. Protected cultivation is a key for sustainable production in the country due to its various ecological regions *viz.*, Western Himalayan region, North eastern Himalayas, Eastern Plateau hill regions, Central Plateau hill regions, Western Plateau hill regions, Southern Plateau hill zone, East coast hill zone, West coast hill regions and Gujarat hill regions (Gupta *et al.*, 2003). Control of light through polycover has become a significant part of horticulture technology because quality, intensity and duration of light is involved in many physiological processes of the plants. A plastic film or fibre glass covering over a green house acts like selective radiation filter which allows solar radiation to pass through but traps radiation emitted by the plants inside the greenhouse which is known as 'Greenhouse effect'. Carbon dioxide released by the plants at night also trapped, this raises the level of CO_2 there by enhances the photosynthetic activity (Ahmed, 1996).

History and Overview of Protected Cultivation in Different Countries

The growing of off-season cucumbers under transparant stone for Emperor Tiberius in the first century,is the earliest reported protected cultivation of horticultural crops. The technology was rarely employed during the next 1500 years. The idea of growing plants in a environment-controlled condition goes back to Roman times. The first modern greenhouse, covered with glass was built in Italy in the 13th century to grow exotic plants that introduced from the tropics. In Asia, China, started protected cultivation in 1990's, and more than 90% of the greenhouses in China are used for vegetable production. The total area covered in China increased to 2.5 m ha under plastic covered greenhouses (85% of the world coverage), 9.6 million ha under plastic mulch and 0.9 m ha underlow tunnels and floating covers.In Europe during winter months underground horse or cow dung hotbeds covered with mica sheets were used for raising leafy vegetables. The greenhouse industryis originated in Holland (1600 AD), during the first half of the seventeenth century (Ambad *et al.*, 2003) In France and England some of the glasshouses were established during 17th century. Development of greenhouse industry in America followed much latter because of its dependence on the economic growth of this land. The first reported green house in America was that of James Neckman in 1764 located in New York. The use of plastic materials such as greenhouse covers is relatively recent. The small countries like Colombia and Kenya had been emerged as quality producers of horticulture crops in proteceted cultivation.Commercial production of vegetables and cut flowers started in the 19thcentury in Europe.In early 19th century glasshouses in different designs came up in Europe and Asian countries, mainly in The Netherlands and Japan, respectively.In the 20th century due to economic development, especially after the Second World War, led to the construction of glass/greenhouses and polyhouses in large areas. In the Western Europe especially Belgium after 2nd world war, it has been characterized by an important development of ornamental plant production under cover and was tripled within ten years

Present scenario of protected cultivation in the World

Country	Area (ha '000 ha)
China	81.0
Spain	70.04
South Korea	47.0
Japan	36.0
Turkey	25.0
India	25.0
Italy	16.5

History and Overview of Protected Cultivation in India

In India protected cultivation is sunrise industry and the area under greenhouse is increasing steadily.India's first exposure to truly hi-tech protected farming of vegetables and other high-value horticultural produce came through the Indo-Israel project on greenhouse cultivation, initiated at IARI, New Delhi during 1998. The last decade of 20th century was marked with globalization and greenhouse industry took firm foundation in India with a reasonable supports from multinationals. Riverbed cucurbits cultivation involving protection of young seedlings from cold winds in winter months by using straw fence is a traditional example in North India. Indo-American Hybrid Seeds (India), Bangalore is the pioneer in India to make use of greenhouse technology since 1965 for commercial productions of flower seeds, cut flowers, ornamental plants andvegetables. National committee on use of plastics in agriculture (NCPA) in 1981 under the Ministry of Chemicals and Petrochemicals further boosted protected cultivation. The NCPA operated through 22 Plasticulture Development Centres (PDC) spread all over the country promoting protected cultivation (use of hail prevention nets, shade nets and packaging of produce)and GATT (General Agreement on Trade and Tariff) promoted the export oriented horticulture. Similarly several protected units in Maharashtra and Karnataka are producing high quality flowers and exporting the same.

Present Status of Protected Cultivation in India

- The states that have consistently expanded the area under protected cultivation are Andhra Pradesh, Gujarat, Maharashtra, Haryana, Punjab, Tamil Nad, Karnataka and West Bengal. Maharashtra and Gujarat had a cumulative area of 5,730.23 hectares and 4,720.72 hectares respectively under the protected cultivation(2015-16). In the coming years, it is expected that the states shall actively engage themselves in promoting protected cultivation in the states through various subsidy schemes and incentive programs for the protected cultivation.

- Thre are **153 units** registered as an export oriented units (EOU) with Gov. of India out of which **65 units** are reported to be producing cut roses which is single highest used flower in the world This area is on the increase due to the efforts made under National Horticulture Mission, Ministry of Agriculture, Government of India

Protected Cultivation in Different States of India

Karnataka: Seed multiplication of flowers and vegetables is practiced and popularized by M/S Indo -American Hybrid Seeds lead by Dr. Manmohan Atawar, is a well known example. This has been adopted in Karnataka by M/S Namdhari Seeds Pvt. Ltd, other seed companies and a large number of farmers. M/s Namdhari Seeds is exporting fresh vegetables grown under green/ polyhouses by farmers with a contract system of vegetable production.

Maharashtra: Early vegetable and flower nursery production under protected structures has made possible cultivation of several vegetables hitherto considered impossible. Floriculture under the greenhouse is confined to Pune, Nasik, Satara, Sangli, Kholapur and Ahmadnagar.There are more than 800 green houses are erectedby the progressive farmers, co-operative and marginal farmers producing cut flowers and vegetables.

Punjab: Net-house vegetable production is becoming popular. Through a systematic economic analysis of net–house, cultivation of vegetables like tomato, brinjal, capsicum, potato, peas, chillies has been found profitable. The gross and net returns were double in net house cultivation than traditional methods of growing of these crops. An early crop of Chappan kaddu (*Cucurbita pepo*) is taken under polyhouses in Punjab which give high return to the farmers. A hi-tech greenhouse covering an area of one acre with capacity to produce up to 35 lakh seedlings of fruits and vegetables at one time has been set up by a farmer with technical expertise from Israel.

Rajasthan: Mr. Sunil Kumar could succeed to produce 5 tons of high quality cucumber from 1000 sq. m. greenhouse with the first crop and 7.5 tons from the second crop and he marketed the entire produce to Delhi niche markets and attained around rupees 4 lakh as gross income from these two crops.

Haryana: It has become a major producer of button mushrooms by adopting black polyethylene protected structures in waste land mainly in Sonipat and Panipat districts. Several spawn production and training centres are being run by NGOs.

Himachal Pradesh: Bestowed with varied agro- climatic conditions, proved a boon to the vegetable growers as natural greenhouse. In 1990-91, the vegetable productivity was hovering around 15.2t/ha. The horizontal expansion in area and introduction of high yielding varieties coupled with refined production technology, enhanced productivity to 18.12t/ha in 2000-01 and 20.00t/ha in 2011-12. Protected cultivation started in 2009 by covering 55ha area, has expanded to 125 ha in 2011-12. The overall productivity in polyhouse is 100t/ha, which is thesole factor motivating the farmers for adoption of poly-culture. The flower production has also appeared on State map by covering an area of 860 ha in open and 70 ha under poly houses.

Scope and Future Thrust of Protected Cultivation in India

- Large scale production and distribution of healthy vegetable and flower seedlings to the large section of growers on nominal price.
- Higher yield per unit area.
- Production of quality horticultural produce.
- Government should support and promote protected cultivation in cluster approach especially in peri-urban areas.
- Development of varieties and hybrids suitable for protected cultivation.
- Promotion of organic production under protected structures.
- Marketing is the key for success of protected cultivation.
- Protected cultivation technology needs to be given a prioritized boost by the State Government.
- Self-constructed greenhouses and low cost temporary structures may also be considered for subsidy linked schemes.
- The quality of structures and technical adherence to dimensions and material used, have to be ensured through multiple-agency verification.
- Promotion of terrace poly gardens.

Constraints of Protected Cultivation

- Initial investment is high.
- Mindset of the farmers to go with the traditional agricultural system and Crops.
- Non availability of adequate post harvest management facilities inthe state.
- Crop varieties/hybrids for protected cultivation are to be imported as there ismeagre research for development of indigenous varieties/ hybrids forprotected cultivation.
- Infrastructure is not available (cold storage, refrigerated vans and others).

- Proper export channels are not established.
- Non availability of markets for products.
- Creation of horticultural estates.
- Quality control measures and farmers organizations.

Conclusion

- Employment generation- both direct and indirect to theunemployed youth and rural unskilled women.
- All kinds of protected cultivation technologies can be adopted in variousareas of the state for cultivation of large number of vegetables, flowers and few fruits.
- Adoption of protected cultivation in peri-urban areas which can build the socio- economic status of farmers.
- Protected cultivation technology is known for increased input use efficiency which is a key to successful farming.
- Plant propagation/seedling production under cover is the need of the day.

References

Ambad, S.N., Kadam, U.S., Dhawale, B.C., Takte, R.L. And Bankar, B.C., 2003, 'Protected cultivation of hortcultural crops. All India Seminar on Potential and Prospects for Protected Cultivation', December 12-13, 2003. At the Institution of Engineers (I), Ahmednagar Local Centre, NHB, Maharashtra.

Balraj Singh, 2012, Protected Cultivation of Vegetable Crops. Kalyani Publishers, Ludhiana, India.

Gupta, R.K., Pandey, R.K. and Sharma, M., 2003, Protective Cultivation- A Key for Sustainable Production in North Western Himalayas, All India Seminar on Potential and Prospects for Protected Cultivation', December 12-13, 2003. At the Institution of Engineers (I), Ahmednagar Local Centre, NHB, Maharashtra.

Nair, R. and Barche, S., 2014, 'Protected Cultivation of Vegetables – Present Status and Future Prospects in India', Indian Journal of Applied Research , (4): 245-247.

Prasad, S. and U. Kumar, 2010, Greenhouse Management for Horticultural Crops. Agrobios (India).

Singh, Brahma. 2012. Protected cultivation of vegetables and flowers-potentials and success stories in India and Abroad. In: "Stakeholders meeting on Protected Cultivation for Haryana", held at Haryana Kisan Ayog, Kisan Bhawan, Khandsa Mandi, Gurgaon on 8th February, 2012.

2

Protected Cultivation

G. Koteswara Rao, T. Thomson, M. M. Nagaraju
M. Surendra Babu

Department of Vegetable Science, Navsari Agricultural University, Navsari Gujarat-396450

Abstract

India is the second largest producer of vegetable crops in the world. However, its vegetable production is much less than the requirement if balanced diet is provided to every individual. There are different ways and means to achieve this target, e.g., bringing additional area under vegetable crops using hybrid seeds, use of improved agro-techniques. Another potential approach is perfection and promotion of protected cultivation of vegetables. In hilly areas parts of the country especially in Northern plains the soils are highly fertile but extremes of temperature ranging from 0-48°C during the year do not allow year round outdoor vegetable cultivation. Similarly, in several parts of the country biotic stresses mainly during rainy & post rainy season do not allow successful production of vegetables like tomato, chilli, okra, cauliflower etc in the fields. In upper reaches of Himalayas, cool desert conditions period where the temperature is extremely low (-5to-30°C) during winter season and most of the region remain cut off from rest of the country from November to March due to very heavy snowfall therefore it is very difficult to grow vegetables in such climate. In many European Countries, USA, Japan, China, Israel, Morocco, Turkey etc where climate reduces the choices for year round outdoor production, vegetables are being produced in protected environments. Greenhouses being the most efficient means to overcome climatic diversity. Greenhouse vegetable production make the use of recent advances in technology to control the environment for maximizing crop productivity percent area and increasing the quality of vegetables produce. India has entered into the area of greenhouse vegetables cultivation more recently and the total area under protected vegetable production is not more than 10,000 hectares. India being a vast country with diverse and extreme agro-climatic conditions, the protected vegetables cultivation technology can be utilized for year round and off-season production of high value, low volume vegetables, crops production of virus free quality seedlings, quality hybrid seed production and as a tool for disease resistance breeding programes.

Introduction

India is a land of diverse agro climatic zones and each of these zones offer a great potential for cultivation of wide range of crops across all seasons. Vegetables form major and important part of our dietary requirements, which are widely grown in the rural and peri-urban areas. Hi-tech horticulture including protected cultivation of high value and exotic vegetables has been on the increase, targeting high end domestic and export market. Of late, due to the population pressure, fragmentation of land holdings and urbanization has led to decline in cultivable area, more so in urban and peri-urban areas. Production of vegetables under protected cultivation system results in effective use of the land resources, besides being able to increase the production of quality vegetables both for the export and domestic markets by offsetting biotic and abiotic stresses to a great extent that otherwise is prevalent in open cultivation. Under protected cultivation, vegetables are widely grown due to higher productivity and economic feasibility.

India is the second largest producer of vegetable crops in the world. However, its vegetable production is much less than the requirement if balanced diet is provided to every individual. There are different ways and means to achieve this target, e.g., bringing additional area under vegetable crops using hybrid seeds, use of improved agro-techniques. Another potential approach is perfection and promotion of protected cultivation of vegetables. In hilly areas parts of the country especially in Northern plains the soils are highly fertile but extremes of temperature ranging from 0-48°C during the year do not allow year round outdoor vegetable cultivation. Similarly, in several parts of the country biotic stresses mainly during rainy & post rainy season do not allow successful production of vegetables like tomato, chilli, okra, cauliflower etc in the fields. In upper reaches of Himalayas, cool desert conditions period where the temperature is extremely low (-5to-300C) during winter season and most of the region remain cut off from rest of the country from November to March due to very heavy snowfall therefore it is very difficult to grow vegetables in such climate. In many European Countries, USA, Japan, China, Israel, Morocco, Turkey etc where climate reduces the choices for year round outdoor production, vegetables are being produced in protected environments. Greenhouses being the most efficient means to overcome climatic diversity. Greenhouse vegetable production make the use of recent advances in technology to control the environment for maximizing crop productivity percent area and increasing the quality of vegetables produce. India has entered into the area of greenhouse vegetables cultivation more recently and the total area under protected vegetable production is not more than 10,000 hectares. India being a vast country with diverse and extreme agro-climatic conditions, the protected vegetables cultivation technology can be utilized for year round and off-season production of high value, low volume vegetables, crops production of virus free quality seedlings, quality hybrid seed production and as a tool for disease resistance breeding programes.

History of Protected Cultivation

The idea of growing plants in environmentally controlled areas has existed since Roman times. The Roman emperor Tiberius ate a cucumber like vegetable daily. The Roman gardeners used artificial methods (similar to the greenhouse system) of growing to have it available for his table every day of the year. Cucumbers were planted in wheeled carts which were put in the sun daily, then taken inside to keep them warm at night. The cucumbers were stored under frames or in cucumber houses glazed with either oiled cloth known as *specularia* or with sheets of selenite (*lapis specularis*), according to the description by Pliny the Elder. In the 13th century, greenhouses were built in Italy to house the exotic plants that explorers brought back from the tropics. They were originally called *giardini botanici* (botanical gardens). Greenhouses in which the temperature could be manually manipulated first appeared in 15th century Korea. The 15th century treatise, the *Sanga Yorok*, contains descriptions of greenhouses designed to regulate the temperature and humidity requirements of plants and crops. One of the earliest records of the Annals of the Joseon Dynasty in 1438 confirms growing mandarin orange trees in a traditional Korean greenhouse during the winter and installing an *ondol* system to provide heat.

The concept of greenhouses also appeared in the Netherlands and then England in the 17th century, along with the plants. Some of these early attempts required enormous amounts of work to close up at night or to winterize. There were serious problems with providing adequate and balanced heat in these early greenhouses. Today, the Netherlands has many of the largest greenhouses in the world, some of them so vast that they are able to produce millions of vegetables every year.

The French botanist Charles Lucien Bonaparte is often credited with building the first practical modern greenhouse in Leiden, Holland, during the 1800s to grow medicinal tropical plants. Originally only on the estates of the rich, the growth of the science of botany caused greenhouses to spread to the universities.

The French called their first greenhouses *orangeries*, since they were used to protect orange trees from freezing. As pineapples became popular, *pineries*, or pineapple pits, were built. Experimentation with the design of greenhouses continued during the 17th century in Europe, as technology produced better glass and construction techniques improved. The greenhouse at the Palace of Versailles was an example of their size and elaborateness; it was more than 150 metres (490 ft) long, 13 metres (43 ft) wide, and 14 metres (46 ft) high.

The golden era of the greenhouse was in England during the Victorian era, where the largest glasshouses yet conceived were constructed, as the wealthy upper class and aspiring botanists competed to build the most elaborate buildings. A good example of this trend is the pioneering Kew Gardens. Joseph Paxton, who had experimented with glass and iron in the creation of large greenhouses as the head gardener at Chatsworth, in Derbyshire, working for the Duke of Devonshire, designed and built The Crystal Palace in London, (although the latter was constructed for both horticultural and non-horticultural exhibition).

Other large greenhouses built in the 19th century included the New York Crystal Palace, Munich's Glaspalast and the Royal Greenhouses of Laeken (1874–1895) for King Leopold II of Belgium. In Japan, the first greenhouse was built in 1880 by Samuel Cocking, a British merchant who exported herbs. In the 20th century, the geodesic dome was added to the many types of greenhouses. Notable examples are the Eden Project, in Cornwall, The Rodale Institute in Pennsylvania, the Climatron at the Missouri Botanical Garden in St. Louis, Missouri, and Toyota Motor Manufacturing Kentucky.

Greenhouse structures adapted in the 1960s when wider sheets of polyethylene film became widely available. Hoop houses were made by several companies and were also frequently made by the growers themselves. Constructed of aluminum extrusions, special galvanized steel tubing, or even just lengths of steel or PVC water pipe, construction costs were greatly reduced. This resulted in many more greenhouses being constructed on smaller farms and garden centers. Polyethylene film durability increased greatly when more effective UV-inhibitors were developed and added in the 1970s; these extended the usable life of the film from one or two years up to 3 and eventually 4 or more years. Gutter-connected greenhouses became more prevalent in the 1980s and 1990s. These greenhouses have two or more bays connected by a common wall, or row of support posts. Heating inputs were reduced as the ratio of floor area to exterior wall area was increased substantially. Gutter-connected greenhouses are now commonly used both in production and in situations where plants are grown and sold to the public as well. Gutter-connected greenhouses are commonly covered with structured polycarbonate materials, or a double layer of polyethylene film with air blown between to provide increased heating efficiencies.

What is Protected Cultivation?

"Protected cultivation practices can be defined as a cropping technique wherein the micro climate surrounding the plant body is controlled partially or fully as per the requirement of the vegetable species grown during their period of growth."

Advantages of Protected Cultivation?

The advantages of protected cultivation are;

- Higher productivity resulting in increased yield,
- Provides better growing environment to plants,
- Protects from rain, wind, high temperatures and minimizes the damage of insect pests and diseases thereby improving the quality and yield,
- Facilitates year round production coupled with yield enhancement by 2-3 times compared to open cultivation.
- Vegetable forcing for domestic consumption and export
- Raising of offseason nurseries
- Productivity is manifold in green houses in comparision to growing the vegetables in open field
- Hybrid seed production
- Maintenance and multiplication of self incompatible line for hybrid seed production
- Polyhouse for plant propagation

What is Greenhouse ?

A greenhouse is a framed or inflated structure covered with transparent or translucent material in which crops could be grown under the condition of at least partially controlled environment and which is large enough to permit persons to within it to carry out cultural operations. The term green-house generally referred to an artificially heated structure. Two or more greenhouses in one location are referred as a greenhouse range.

Working Principle of Greenhouse ?

The correct explanation to how does a greenhouse work is that it starts with the heat carrying rays of the sun, entering the walls and/or roof of the greenhouse and heating everything within. The temperature of the ground rises as it absorbs this radiation. The heat from the ground is transmitted to the layer of air next to it, which expands and becomes lighter than the air above it. This heated lighter air rises and is instantly replaced by cooler denser air. This heating cycle continues through the day. The roof and the walls of the greenhouse keep in this heated air, and thus the air in the greenhouse stays warm all day long. In an open space, the heating of air is spread over a large mass, and is thus diluted. In the nights, the temperature in a greenhouse stays warmer that the air outside as heat stored during the day is available through the night.

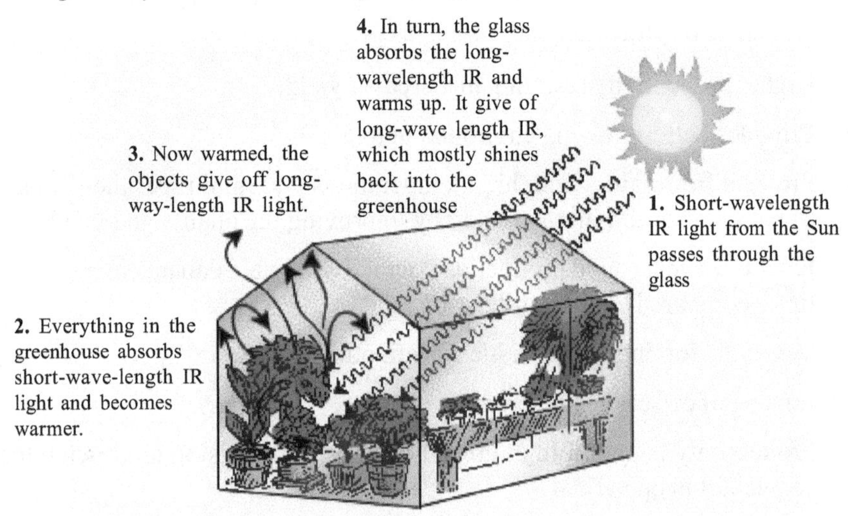

Uses of Greenhouses ?

- Under greenhouse one can grow crops controlled environment and throughout the year four or five crops can be grown due to availability of required plant environment conditions. This helps in increased productivity and superior quality of produce can be obtained.
- Gadgets for efficient use of various inputs like water, fertilizer, seed and plant protection chemicals can be well maintained in greenhouse.
- Pests and diseases are effectively controlled in the growing enclosed area.
- Percentage of germination of seeds is high in greenhouse.
- The acclimatization of tissue culture plantlets can be carried out in greenhouses.
- Agricultural and horticultural crops productions schedules can be planned to take advantages of market need.
- Different types of growing media like peat mass, vermiculate, rice hull and compost that are used in intensive agriculture can be efficiently used in greenhouses.

- Export quality produce of international standard can be produced by utilizing the entrapped heat.
- When the crops are not grown, drying and related operations of the harvested produce can be taken up utilizing the entrapped heat.
- Greenhouses are suitable for automation of irrigation, application of other inputs and environment control by using computer and artificial intelligence techniques.
- Self-employment for educated youth on farm can be increased.

Crops Covered under Greenhouse Conditions

The following crops are most widely commercially cultivated under green house conditions they are capsicum, tomato, cucumber, cabbage, strawberry, gerbera, carnation and rose etc.

Greenhouse Construction

The structural components of greenhouse includes

Sidewall: supports the entire weight of the greenhouse. Consists of the following parts -

- side ventilators
- side posts
- curtain wall

Concrete footer: foundation for the side posts

Eave: area where the sidewall joints connect

Sash bar: bar made of metal or wood that runs from eave to the ridge, holding glass in place.

Trusses : added to framework for structural support Composed of rafters chords struts

Purlins: runs the length of the greenhouse and are bolted to each truss.

Ridge: top of the greenhouse.

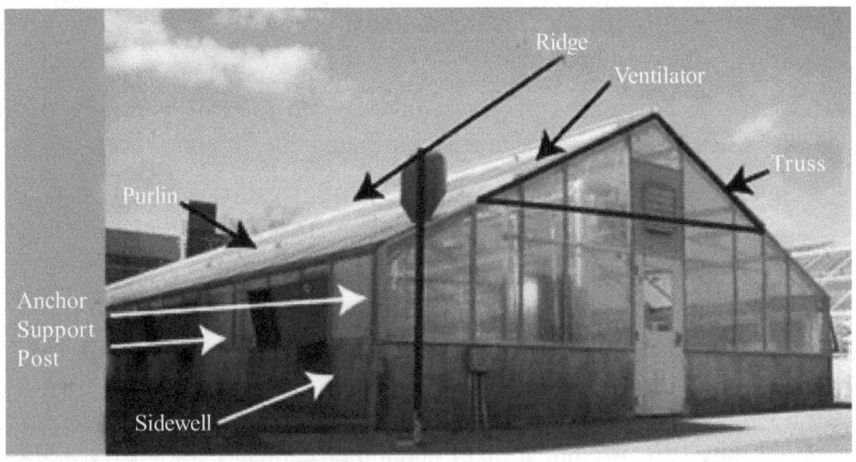

Basic designs of Greenhouse

There are three specific designs of greenhouses. They are

I. Attached greenhouse
II. Free standing greenhouse
III. Connected greenhouses

Types of Greenhouses

Greenhouses are a technology based investment. The higher the level of technology used the greater potential for achieving tightly controlled growing conditions. This capacity to tightly control the conditions in which the crop is grown is strongly related to the health and productivity of the crop. The following three categories of greenhouse have been defined to assist people in selecting the most appropriate investment for their needs and budget.

Low Technology Greenhouses

These greenhouses are less than 3 m in total height. Tunnel houses, are the most common type. They do not have vertical walls. They have poor ventilation. This type of structure is relatively inexpensive and easy to erect. Little or no automation is used. While this sort of structure provides basic advantages over field production, crop potential is still limited by the growing environment and crop management is relatively difficult. Low level greenhouses generally result in a suboptimal growing environment which restricts yields and does little to reduce the incidence of pests and diseases. Pest and disease control, as a result, is normally structured around a chemical spray program. Low technology greenhouses have significant production and environmental limitations, but they offer a cost effective entry to the industry.

Medium Technology Greenhouses

Medium level greenhouses are typically characterized by vertical walls more than 2m but less than 4 m tall and a total height usually less than 5.5 m. They may have roof or side wall ventilation or both. Medium level greenhouses are usually clad with either single or double skin plastic film or glass and use varying degrees of automation.

Medium level greenhouses offer a compromise between cost and productivity and represent a reasonable economic and environmental basis for the industry. Production in medium level greenhouses can be more efficient than field production. Hydroponic systems increase the efficiency of water use. There is greater opportunity to use non-chemical pest and disease management strategies but overall the full potential of greenhouse horticulture is difficult to attain.

High Technology Greenhouses

High level greenhouses have a wall height of at least 4 m with the roof peak being up to 8 m above ground level. These structures offer superior crop and environmental performance. High technology structures will have roof ventilation and may also have side wall vents. Cladding may be plastic film (single or double), polycarbonate sheeting or glass. Environmental controls are almost always automated. These structures offer enormous opportunities for economic and environmental sustainability. Use of pesticides can be significantly reduced. High technology structures provide a generally impressive sight and, internationally, are increasingly being involved in agribusiness opportunities. Although these greenhouses are capital intensive, they offer a highly productive, environmentally sustainable opportunity for an advanced fresh produce industry. Investment decisions should, wherever possible, look to install high technology greenhouses.

Greenhouses vary in style, size and materials that are used to build it in order to fulfill any requirements and to suit any type of crop. The materials used to build the main structure of a greenhouse are timber, aluminum or steel. Timber frames are the traditional choice for garden greenhouses and hardwoods require low maintenance. Aluminum alloy frames are more lightly, need only minimum maintenance but are extremely sturdy. Steel frames are very strong but must be treated regularly to prevent them from rust, but they are also cheaper than timber or aluminum frames. For glazing you can use glass or plastic panels. Size may also vary according to your necessities.

Many styles of greenhouses are available on the market, every one of them with specific qualities: some provide optimum ventilation, or best use of space, or conserve heat well or allow better light penetration but all of them are made in order to fulfill your personal preferences.

Greenhouses types can be split into two main categories: conventional greenhouses and specialist greenhouses. Conventional greenhouses include: traditional span, Dutch light, three-quarter span, lean-to and mansard or curvilinear greenhouses. Specialist greenhouses include: Dome-shaped, polygonal, alpine house, conservation, mini and poly-tunnel greenhouses.

In the next lines we will try to describe every of those types and see what are their main qualities to help you choose the most suitable greenhouse for your garden.

Traditional Span

This type of greenhouse is practical in terms of growing space and headroom by its vertical sides and even span roof. It provides the best use of space for the least cost for raising seedlings and growing border crops. Its lower part stop the heat lost over the winter.

Dutch Light

This type of greenhouse is designed in order to allow in maximum light through the sloping sides. It is suitable to grow border crops, preferably low-growing ones. The panes of glass on the roof overlap slightly to keep out rain but also to increase the rigidity of the structure.

Three-quarter Span

This type of greenhouse is positioned with one of its sides against a wall, preferably beside a sunny wall because the light is a little more restricted than in a free-standing greenhouse, but this also mean that it will need some extra shading in the summer. If you will choose a house wall to position your greenhouse you will also benefit from extra warmth and insulation from this.

Lean-to Lean

One can use this type of greenhouse in a garden with insufficient space for a free-standing structure. Like the three-quarter span, this type of greenhouse will benefit from the warmth and insulation of the house wall. Many of those greenhouses are similar in appearance to conservatories and may be used as garden rooms. In this type of greenhouse installation of electricity, gas or water supply easier and cheaper than that of a greenhouse located at some distance from the house.

Mansard or Curvilinear

This greenhouse has slanting sides and roof panels designed to allow in maximum light available so a best place for this type of greenhouse is an open site with no shade from the surrounding trees or buildings. This greenhouse is suitable for plants that need maximum light over the winter.

Dome-shaped

This type of greenhouse is offering an elegant design that is mostly useful in exposed positions. It is stable and offers less wind resistance than traditional greenhouses. It allows maximum light transmission because of its multi-angled glass panels. It might offer limited headroom around the edge.

Polygonal

For a focal point in the garden or for gardens where appearance is important those greenhouses are the most used. Any octagonal or polygonal greenhouse is a good choice, but they may be more expensive than traditional greenhouses of similar sizes.

Alpine House

Traditionally those greenhouses have timber-frame with louvre vents all along the sides. This help for most effective ventilation. Usually, these types of greenhouses are not heated and they are not closed unless the winter is too cold, so the insulation is not needed. They are used mostly for plants that just need some protection from dampness and rain and require a bright and well-ventilated place.

Conservation

This type of greenhouse is designed to save as much energy as possible using special features. The roof panels are angled to permit optimum light penetration. Mirrored surfaces are also used to reflect light within the greenhouse itself. With all those special features, this type of greenhouse is usually more expensive than others of the same size.

Mini

For a limited space in your garden, or if you only have a small number of plants to grow, this useful, low-cost greenhouse is available in different sizes and also as free-standing or wheeled versions. Made from aluminum frame and covered with plastic or glass, this greenhouse is best to be placed face SE or SW in order to get the maximum light penetration. Access may be a problem as all the work has to be done from the outside. Venting and shading in the summer are essential.

Poly-tunnel

For a low-cost protection, for the vegetable plot for example, a plastic poly-tunnel greenhouse is the choice. It covers a large area, is covered with heavy-duty transparent plastic sheets and offers protection from cold and wind, is easy to move where needed, so is the perfect choice for your crops, either you choose to plant them directly into the soil, in pots or in growing bags. Ventilation may be a problem and the sheets may need to be replaced every few years as they gradually become opaque.

Basis for Classification of Greenhouse

Greenhouse structures of various types are used for crop production. Although there are advantages in each type for a particular application, in general there is no single type greenhouse, which can be said as the best. Different types of greenhouses are designed to meet the specific needs. The different types of greenhouses are classified based on shape, utility, material and construction are given below:

1. Greenhouse type based on shape

The types of greenhouses based on shape are :

a) Lean to type greenhouse
b) Even span type greenhouse
c) Uneven span type greenhouse
d) Ridge and furrow type
e) Saw tooth type
f) Quonset greenhouse
g) Interlocking ridges and furrow type Quonset greenhouse
h) Ground to ground greenhouse.

2. Greenhouse type based on utility

Greenhouse classification can be made depending on the functions or utilities. Among the different utilities, artificial cooling and heating are more expensive and elaborate. Hence based on this, they are classified into two types.

a) Greenhouses for active heating
b) Greenhouses for active cooling

3. Greenhouse type based on construction

The type of construction predominantly is influenced by structural material, though the covering material also influences the type. Higher the span, stronger should be the material and more structural members are used to make sturdy tissues. For smaller spans, simple designs like hoops can be followed. So based on construction, greenhouses can be classified as

Lean type greenhouses

Even span or gable type greenhouse

Attached uneven span type

Saw tooth type greenhouses

Quonset type greenhouses

Gothic greenhouses

Fig. 1 : Classification of Greenhouses based on shape

Wooden framed greenhouse

Pipe framed greenhouse

Truss framed greenhouse

Fig. 2 : Classification of greenhouses based on construction

Glass greenhouse

Plastic greenhouse

Fig. 3 : Classification of greenhouses based on cladding material

a) Wooden framed structure
b) Pipe framed structure
c) Truss framed structure

4. Greenhouse type based on covering material

Covering materials are the important component of the greenhouse structure. They have direct influence on greenhouse effect, inside the structure and they alter the air temperature inside. The types of frames and method of fixing also varies with covering material. Hence based on the type of covering material they may be classified as

a) Glass glazing
b) Fibre glass reinforced plastic (FRP) glazing
   i) Plain sheet
   ii) Corrugated sheet.
c) Plastic film
   i) UV stabilized LDPE film.
   ii) Silpaulin type sheet.
   iii) Net house.
d) Based on the cost of construction involved
   i) High cost greenhouse
   ii) Medium cost greenhouse
   iii) Low cost greenhouse

Constraints in Protected Vegetable Production

In India polyhouse culture is in infant stage and has not become popular as yet. High cost and non-availability of various components and small scale farmers are the major limiting factors in the adoption of polyhouse technology for commercial cultivation. Many of the polyhouse components like fiber glass, cooling pads, fans, etc have to be imported at high costs including freight and custom duty. Greenhouse and other structures design for different agro-climatic of the region is not standardized. Lack of awareness among farmers pertaining to potentials of protected vegetable production and lack of major research programme on protected vegetable farming are other limiting factors.

Conclusion

The greenhouse technology is still in its preliminary stage in the country and concerted efforts are required from all concerned agencies to bring it at par with the global standards. Economically viable and technologically feasible greenhouse technology suitable for the Indian agro-climatic and geographical conditions is needed at the earliest. Work should be channelised in finding suitable and locally available construction material for low- and medium-cost greenhouses. Utilisation of the solar energy stored in the solar photovoltaic cell or else the heat rejected from the steam turbines of the thermal power plants for greenhouse heating and humidity control needs to be improved. As the area under greenhouse houses in India is likely to increase in the near future, the concerned industries should make earnest efforts to provide the much needed support in the hardware and software fields. Government initiatives/efforts in popularizing the greenhouse technology among the farming community of the country are to be strengthened. There is a need for the Government to encourage the farmers by providing timely subsidy for taking up this new technology in a big way.

References

Anonymous, 2002. Indian Council of Agricultural Research 2002. Agricultural Research Data Book, ICAR, 2004.

Diver, S. Organic greenhouse vegetable production horticulture systems guide: NCAT Agriculture Specialists, ATTRA Publication No. IP078, 2000.

Ghosh, A. (2009). Greenhouse Technology (The Future Concept of Horticulture). Kalyani Publishers, New Delhi.

Gyan P. Mishra, Narendra Singh, Hitesh Kumar and Shashi Bala Singh. 2010. Protected Cultivation for Food and Nutritional Security at Ladakh. Defence Science Journal. 61 (2): -219-225.

Good Agricultural Practices for greenhouse vegetable crops. Principles for Mediterranean climate areas. FAO plant production and protection paper 217.

http://www.greenzonelife.com/greenhouses/greenhouses-types.html

https://courses.worldcampus.psu.edu/welcome/hort101/print.html

Indian Institute of Horticultural Research Technical Bulletin for Capsicum Cultivation. Technical Bulletin number-22 (Revised edition).

Indian Soc. Veg. Sci. Souvenir: Silver Jubilee, National Symposium Dec. 12-14, 1998, Varanasi, U.P. India pp. 90.

Kouser Parveen Wani, Pradeep Kumar Singh, Asima Amin, Faheema Mushtaq and Zahoor Ahmad Dar. (2011). Protected cultivation of tomato, capsicum and cucumber under kashmir valley conditions. Asian Journal of Science and Technology.1(4):056-061.

Prasad, S. and Kumar, U. (2003). Greenhouse Management for Horticultural Crops. Agrobios (India), Jodhpur

Rai, N; Nath, A; Yadav, D.S. and Patel, K.K. 2004. Effect of polyhouse on shelf-life of bell pepper grown in Meghalaya. National Seminar on Diversification of Agriculture through Horticultural Crops, held at IARI Regional Station, Karnal, from 21-23rd February, pp. S.P.22.

Singh; Brahma. 1998. Vegetable production under protected conditions: Problems and Prospects.

Singh, Narender; Diwedi, S.K. and Paljor, Elli. 1999. Ladakh Mein Sabjion Kei Sanrakshit Kheti.

Tiwari, G.N. (2003). Greenhouse Technology for Controlled Environment. Narosa Publishing House, New Delhi.

www.gbcgroup.co.uk/manf/alton/www.agriinfo.in

3

Protected Cultivation Techniques of Horticulture Crops

K. Shoba Thingalmaniyan and N.A. Tamil Selvi

Department of Vegetable Crops, Horticultural College and Research Institute Tamil Nadu Agricultural University, Coimbatore – 3 Tamil Nadu

India is blessed with wide range of diverse agro climatic conditions, making it an appropriate place for vegetable cultivation, resulting in increasing vegetable production from has increased from 63.8 MT in 1993 to 168.5 million tonnes in 2015-16 and productivity from 12.6 MT/ha to 17.8 MT/ha (www.indiastat.com). Although the production has increased but still the technology used and practices followed is predominantly traditional resulting in low productivity and inconsistent quality and quantity of produce supplies to various markets in the country. The per capita consumption of vegetables in India is very low against WHO standards and FAO recommendation (135 g/day <180 g/day (WHO standards) < 300 g/day capita (FAO recommendation).

There are different ways and means to achieve this target, e.g., bringing additional area under vegetable crops using hybrid seeds, use of improved agro techniques. Another potential approach is perfection and promotion of protected cultivation of vegetables. Protected cultivation is the method to protect the crops from consistent fluctuation in rainfall, frequent droughts and infestation of crops by pest and diseases. It is done through greenhouse, mulching, plastic tunnels and shade nets and anti bird/hail. Protected cultivation is being widely used in countries such as Netherlands, Japan, France, Germany and USA. In India the protected cultivation has been implemented in 10000 hectares in 2012. Protected cultivation technology can be utilized for year round and off season production of high value low volume vegetable crops, production of virus free high quality seedlings, quality hybrid seed production and as a tool for disease resistance breeding programmes. The vegetables like tomato, capsicum and cucumber were commercially cultivated under protected structures.

The main factors that can be attributed to protective cultivation methods in India are

- Manifold increase in the agricultural production from the same unit area
- Optimum consumption of water and electricity as compared the conventional cultivation
- Cultivation is possible under unfavorable agro-climatic conditions
- Certain vegetable crops can be grown round the year in a particular place
- Better quality of the produce
- Higher input use efficiency
- Reduction in loss due to pest, disease and weeds
- Income per unit area is high
- Provides excellent opportunities for export
- Early maturity of the crops as compared to conventional cultivation system

Protected Structures used for Greenhouse Cultivation

Vegetable cultivation needs protection from strong wind, low temperature, intense sunshine, rain and hailstorms depending upon the climate of the region. Protection from wind and low temperature is provided by erecting wind breaks with different devices. The style of protected structures for crop production ranges from simple provisions such as rain shelters, shade houses, mulches, row covers, low tunnels, cloches to greenhouse structures with passive or active climate control. The commonly used protected structures are described hereunder.

i. Net houses

These are of two types of simple structures *viz.,* shade nets and insect proof nets. Shade net are perforated plastic materials used to cut down the solar radiation to protect the leaves from wilting/scorching. These nets are available in three colours *viz.,* black, green and white with different shading intensities ranging from 25-75 %. Leafy vegetables and ornamental greens are preferably grown under shade nets especially when sunlight is stronger. Theses are manufactured from high density polyethylene (HDPE) plastics but nowadays Ultra violet (UV) stabilized nets are also available which have longer life.

Insect proof nylon nets are available in different intensities of perforations, ranging from 20-60 mesh. Nets of 40 and above mesh size are effective to control most of the flying insects and save the crop from viral diseases. However, it reduces the air exchange. Early planting of tomato and capsicum is possible under this structure without the risk of vector.

ii. Plastic low tunnels/ row covers

These structures are laid in open fields to cover rows of plants with transparent plastic film stretched over steel hoops of about 50 cm height spaced suitably along the rows. Polyethylene films of 30-40 micron thickness without UV stabilization were used. Row covers were used for vegetable production under temperate and tropical regions. In temperate region, row covers were used to increase soil temperature, stimulate germination, early growth, to protect the plants from frost injury and improve the quality of the crops. In north India, row covers were used for off season production of cucurbits during winter season. Muskmelon cultivation under this row covers showed highly profitable returns. Muskmelon seedlings could be transplanted under such covers in the last week of January. The crop growth was sustained during the winter with the help of row covers. The temperature profile varied around 7°C over 24 hours cycle.

In summer the materials used as row covers need to have adequate permeability to air and moisture to prevent the accumulation of excessive heat inside the covers. The covering materials used in summer are woven polyester wind-break nets, cheese-cloth and insect proof screens. These type of covers are generally laid over the plant rows with any support which is known as floating row covers.

iii. Plastic mulch

Plastic mulching is a practice of covering the surface around the plant to make the conducive climate for plant growth through *in situ* moisture conservation, weed control, better CO_2 exchange for root system and soil structure maintenance. It avoids the contact of fruit surface with the soil by the way it ensures the cleanliness of the produce. Silver and yellow coloured films are successfully used for repelling aphids and whitefly. Low density polyethylene (LDPE) and Linear low density polyethylene (LLDPE) plastic films are commonly used for mulching. Black plastic mulches are commonly used plastic films which check the weed growth.

iv. Trench cultivation

It is a simple and cheap structure for vegetable production under extreme winter even at 15°C temperature. It harness soil and sun heat to create conducive climate for cultivation of certain leafy vegetables. For large scale production, trenches of different length can be made. For holding plastic sheets, wooden poles or metal pipes are used.

v. Floating plastic covers

Transparent plastic sheet is used to cover large open field to protect vegetables from frost/ snow and low temperature.

vi. Greenhouses

Greenhouses are framed structures of variable size, shape, above ground or semi-underground covered with transparent or translucent materials in which crops could be grown under partially controlled environment, which is large enough to permit normal cultural operation. In greenhouses, the growing environment is altered to suit the specific requirement of the plants. It is rather used to protect the plants from the adverse climatic conditions by providing optimum conditions of light, temperature, humidity, CO_2 and air circulation for the best growth of the plants to achieve maximum yield of best quality. It is the most intensive form of commercial cultivation and also called as food factories.

Classification of Greenhouses

a. Classification based on cladding material

Greenhouses are covered with transparent or translucent material either fibre glass or ultraviolet stable polyethylene film. Based on the type of cladding material used in covering the installed structures, greenhouses can be broadly grouped into the following categories.

- Fibre-glass greenhouses
- Single or double polyethylene film greenhouses
- Ordinary glasshouse
- Polycarbonate houses and
- Ultraviolet-stable polyethylene film house

b. Classification based on structure

It is based on the material used for preparation of frame. The material may be wood (wood house), bamboo (bamboo greenhouse) or steel (steel greenhouse). Bamboo framed structures are cost effective for the places where bamboo is locally grown. However, wooden structures have termite problem.

c. Classification based on cost

Cost required for constructing one square meter of greenhouse may vary from low to high depending on the material used and facilities provided.

i. *Low cost greenhouse*

The low cost greenhouse is made of a supporting structure of G.I pipe, angle iron, steel tubing or bamboo. For covering ultraviolet stabilized plastic film of 200 micron thickness is used. It does not have any control system.

ii. *Medium cost greenhouse*

The medium cost greenhouses have cooling and heating arrangements and may have a double layer or ultraviolet stabilized plastic film.

iii. *High cost greenhouses*

It has many facilities like auto control mechanism of heating, cooling and humidification system, drip irrigation system.

d. Classification based on climate control

i. *Hi tech or climate controlled greenhouse*

This type of greenhouse is constructed to achieve higher degree of climate control to enhance the cultivation period. Evaporative cooling and heaters are used to maintain required temperature inside the greenhouse. The climate is usually accomplished through automation system. Such type of greenhouses are used for growing high value crops like tomatoes, cherry tomatoes, sweet peppers and cucumbers for long period. These types of greenhouses are mainly used in several European countries, Morocco, Israel, USA, Japan etc.

ii. *Semi-climate controlled greenhouses*

In this structure, the structural frames are made up of G.I pipes, like climatic controlled greenhouses but only exhaust fans with evaporative cooling pads are provided to maintain favorable temperature and humidity during summer. These types of greenhouses are suitable for vegetable cultivation during mid winters and mild summers only. The basic cost of erection of these greenhouses is half of the climate controlled greenhouses. These greenhouses mainly used in Turkey, Spain, USA, Morocco, Israel and Japan.

iii. *Naturally ventilated or low cost greenhouses*

These are simple construction greenhouses with low initial cost. The frames may be of G.I. pipes, wooden logs or steel pipes but no heating and cooling system are provided for the structure. The top of the greenhouses are provided with plastic whereas the side walls have insect proof net from ground up to 5-6 feet height with or without manually rollable plastic over it. The initial cost of these greenhouses is less than half to that semi-climate controlled greenhouses. These type of greenhouses are mainly used for crops like cucumber, off-season production of muskmelon, tomato etc. These type of greenhouses are used in Japan, Israel, Turkey, Morocco.

Factors Affecting Construction of a Greenhouse

Careful planning prior to construction of greenhouses is the first step in the development of a successful and profitable greenhouse production system. For constructing greenhouses the following factors should be taken into account.

Climate

Greenhouse construction is strongly depending on climate and weather. Excessive temperature in dry areas can seriously limit the growth. Latitude affect the day length, therefore the regions greater than 40° N or S latitude will have significant variations in total solar energy regardless of local conditions such as clouds, obstructions or pollution. Similarly the average mean temperature will change with season.

Location

The ideal location for green house should have high winter light intensity, moderate winter temperature, low humidity and easy access to markets. The easy availability of existing utilities will help to reduce the establishment cost and will affect the ongoing fuel cost. Greenhouses should not be erected near trees or building which shades the greenhouses. Windbreaks will help to reduce the heating cost. These should be located near the big cities for easy transport of the produce.

The diurnal changes in temperature have an effect upon vegetation and greenhouse practices in terms of heating and cooling costs, protection from weather damage and requirements for advanced control systems.

Site Selection

Selection of proper operational site should be the first and foremost criterion for which the following factors must be considered.

- Current and projected land values
- The site should be nearer to a power source for continuous availability of electricity
- It should be nearer to the water source
- Availability of skilled, trainable, reliable manpower at affordable wages
- The sites should have adequate means of communication such as telephone, fax, e-mail connectivity

The site should also be evaluated for its environmental qualities, drainage and soil characteristics.

Cost of Construction

Depending upon the local climatic conditions, materials used, facilities provided and choice of the crop, the initial investment could range from Rs. 125- 2000 per square meter of floor area for a steel frame greenhouse. Material used to construct a greenhouse frame may be wood, bamboo, steel or even aluminum. The wood and bamboo framed structures cost effective for the places where they are locally available. However the wooden structures, run the risk of termite attack.

In indigenous greenhouses, local materials were used as frames and for covering transparent or translucent films and these structures are comparatively quite cheap, but their life is short.

Size of Greenhouse

A retail greenhouse is, generally up to 1000 m^2 growing area. It may be as small as 20 m^2 for specialized raising of seedlings or hardening of tissue cultured plants. The size of greenhouse largely depends upon the available resources and demand of market. The structural design of greenhouse is greatly influenced by the size of greenhouse. A greenhouse up to 250 m^2 area could be a single span structure. However, a larger facility could either be a multi-span or a cluster of several single span units.

Vegetables to be Grown

Crops like tomato, capsicum, cucumber have been found to be more remunerative. The main consideration is choosing the vegetables to be grown inside the limited and expensive space of the greenhouse is the most efficient and economical utilization of space for the longest period possible. In this regard, low volume and high value crops may be the most profitable to grow in the environmentally controlled greenhouse.

Materials Required for Greenhouse Fabrication

a. Frame materials

Depending upon the local availability, the green house frame can be framed with wood, bamboo, bricks, stone, concrete, aluminum, steel etc. The choice of frame material affect the size, shape, environmental control possibility and automation of greenhouse operations. In hilly regions, the locally prepared bricks are used for the east, north and west side of the greenhouse. Only south side is glazed with UV stabilized polyethylene. Bamboo green house are cost effective, good rain shelters and have naturally ventilated facilities. Steel and aluminum framed greenhouses can be prepared with any shape and size. At present galvanized steel is the only viable frame material in India except where good quality bamboo and wood is locally available.

Wood

This is commonly based on availability and where the cost of wood is comparatively low. The construction of structure is simple as local artisans are able to complete the job and, moreover it is easy to construct greenhouse with wood structures.

Steel

The construction of greenhouse with high tensile strength pipe and tubular steel pipes and zinc plate bolts and nuts for anchorage are widely used. There has been considerable development in construction of greenhouse structure with prefabricated steel structures, which are commercially available. Galvanized steel or zinc coated pipes are preferred to avoid corrosion.

Aluminum

It is light in weight, easy to handle and is not adversely affected by most greenhouse constructions. Aluminum pipes, angles, channels and T sections are commercially available. These can be used for trusses, purlins and columns.

b. Glazing materials

A variety of plastic materials are available for greenhouse glazing, *e.g.* UV stabilized, LDPE, PVC, EVA, acrylic, fibre glass, polycarbonate, etc. Plastic glazed greenhouses have several advantages over glass greenhouses. Plastic is also adopted for various greenhouse designs as it is generally resistant to breakage, light weight and relatively easy to apply. An ideal glazing material should have the following characters.

- Transmit the visible portion of sunlight
- It should be highly transparent to absorb the ultraviolet portion and convert some of it into visible light
- Reflect or absorb infra red radiation
- It should have long life
- It should be resistant to weathering and breakage
- It should be cost effective

c. Covering material

Plastic is advantageous than glass because of its lightness and corresponding lighter structure, mobility, flexibility, impermeability, resistance to hail, simple to work, resistant to wind load and lower cost. The important plastics used to cover greenhouses are described hereunder.

(i) *Acrylic*

It is resistant to weathering and breakage and is very transparent. Its ultraviolet radiation absorption rate is higher than glass. Double layer acrylic transmit about 83 per cent of light and reduces heat loss by 20-40 per cent over single layer. This material does not turn yellow. These have disadvantages *i.e.* flammable, very expensive and easily scratched.

(ii) *Fibre glass*

The fibre glass reinforced polyester (FRP) panels are durable, attractive and moderately priced. As compared to glass, FRP panels are more resistant to impact, transmit slightly less light and weathering over time reduces light transmission. This plastic is easy to cut and comes in corrugate or flat panels. Fibre glass has a high expansion/ contraction rate.

(iii) *Polycarbonate*

It is capable of resisting impact better besides being flexible, thinner and less expensive than acrylic. Double layer polycarbonate transmits about 75-80 % of light and reduces heat loss 40 % over single layer. This material scratches easily and has a high rate of expansion/ contraction. Its starts turning yellow and loosing transparency within a year.

Films

a) Polyethylene film

It is inexpensive but temporary, less attractive and requires more maintenance than other plastics. It is easily destroyed by ultraviolet radiation from the sun, although films treated with UV inhibitors will last 12-24 months longer than untreated. Polyethylene film is produced in wider sheet requiring fewer structural framing members for support resulting in greater light transmission. In two layers system reduces heat loss by 30-40 per cent and transmits 75-87 % of available light when new. It is most widely used as a greenhouse film.

b) Polyvinyl chloride film

PVC film has high emissivity for long wave radiation which creates slightly higher air temperature in the greenhouse at night. UV inhibitors can increase the life of film. It is more expensive than polyethylene film and tends to accumulate dirt, which must be washed off during winter for better light transmission.

c) Polyester

It is widely used in textile industry. It has low light transmissivity compare to glass. Polyester film is very strong and can last up to 15 years. The cost is about three to four times higher than that of polyethylene.

Greenhouse Designs

A greenhouse is primarily a plant growing structure covered with transparent material such as glass or polyethylene to let the necessary light pass through necessary for plant growth and suitable to increase the quality of crops raised. The important point for designing a greenhouse is that the structure should be able to transmit maximum possible sunlight during winter season. The basic structure of a greenhouse consists of a sturdy frame capable of resisting wind and snow load. The present day greenhouse technology is although about two centuries old but enough efforts have been made to develop designs for different conditions. As a result designs are available from simple structures to compete environment controlled structures. The major consideration is the design of greenhouse has been the design of shape of the roofs. Conventional foundations usually support vertical walls and the roof may be glazed, trussed or arched. The conventional greenhouse is fitted with glass but plastic film, fiberglass panels and polycarbonate panels often supplement glass. The design should be such that angle of incidence of solar radiation is never greater than 40.

The structural design of greenhouses must provide for safety from wind, snow and many other forces while permitting adequate light transmission for the crop. The opaque and many other forces permitting adequate light transmission for the crop. The framing members should not provide undue shading and yet provide sufficient strength to resist anticipated loads over the expressed life of the structure. Greenhouses are normally designed for a service life of 25 years. These are light weight structure and seldom exceed 9 m in height. The problem is more often to hold them down than to keep them up. Standards for construction vary with country, although the engineering of such buildings remain more or less the same.

In addition to bearing the weight of the structures, greenhouses are subjected to a number of other stresses that must be taken into account. In addition to the dead load, the designer must take into consideration the live loads, wind loads and snow loads to develop a structurally sound and economical useful greenhouse. Wind load in parts of the world may be more important than snow loads. The greenhouse structures are usually designed to withstand the speed of at least 130km / hour. There are several styles of commercial greenhouse designs have been developed. The mansard and uneven span types are not common due to high cost. Saw tooth types are common in mild climate. The Venlo type is considered as probably the standard design at least in European countries. The very wide spans such as the vinery and mansard are less common.

In any event, attempting to design structures for all possible events leads to an expensive and insufficient building. The essence of good design is to build for most probable conditions that will be encountered providing an efficient greenhouse at minimal cost. In such cases, some risk will always be present that cannot be avoided.

The essence of good design is to build for most probable conditions that will be encountered providing an efficient greenhouse at minimal cost. In such cases some risk will always be present that cannot be avoided. The owner in cooperation with the manufacturer must access those risks realistically assigning a probability factor to serve weather. On this basis, the grower can determine a risk factor for a particular structure. For avoiding the risk in the designing of a greenhouse structure, the design load calculations should be done carefully.

Design Load

It is essential to consider all the loads while erecting a greenhouse. The design of greenhouse structure is mainly governed by the dead load, live load and snow load.

- Dead load is the weight of materials such as frame made up of GI pipes or wooden structure, weight of polythene sheet etc. were used in the construction of greenhouse such as floor, roof, framing and covering
- Live load is the weight superimposed by use (like baskets, shelves attached, vine ropes, etc) but not wind, snow and dead load
- Snow load is the vertical load applied to the horizontal projection of the roof
- Wind load is caused by blowing wind in any horizontal direction *i.e.* load due to wind velocity

Structural Components

Structural members of the greenhouses are hoops, foundation, lateral support, polygrip assembly and end frames. Details of 4 m x 24 m Quonset greenhouses are given below

1. **Hoops**

   Hoops are integral part of greenhouse frame. These are semi-circular in shape and are formed using bending galvanized iron pipe (15 mm diameter and 5.9 m length). As discussed earlier, pipe blender should be used to give pipe curvature to the pipe.

2. **Foundation pipe**

   Foundation pipes are meant to provide a firm support to the hoops and to provide the polygrip firmly. For foundation pipes generally 90 cm long pieces of 25 mm diameter GI pipe (Glass A) are used. A 10 cm piece of mild steel (MS) flat (25 mm x 3 mm) is centrally welded to the one end of the pipe and 9 hole of 8 mm diameter drilled at 10 cm distance from the other end. The flat welded end is fixed in the hole to a depth of 75 cm grounding with concrete. Whole hoops and polygrip assembly along with 15 cm pipe is kept above the ground level. The foundation pipes are spaced 1 m apart in parallel rows. A special care is taken to ensure that the top of all foundation pipes are at the same elevation.

3. **End frame**

   End frames are wooden members filled on both ends of the greenhouse. The end frame structures are made out of 5 x 5 cm section. This member has the provision for door and supporting fixtures for the environment control equipments.

4. **Lateral support**

   Lateral supports are provided to exchange the structural rigidity of end frames and are fabricated from 10 mm diameter mild steel rods. A ring of 3.5 cm diameter is made at the one end, and a right angle hook at the other end.

5. **Polygrip assembly**

   The polyethylene covering the greenhouse is firmly fixed with the foundation pipe to ensure rigidity against wind load without being blown off. The polygrip mechanism has been designed in such a way that while it holds polyethylene firmly, puncturing avoided. It is made up of 20 gauge galvanized iron sheet. The polyethylene sheet is stretched and mild steel pieces are put at a distance of 50 cm, holding the right angle strip against the channel along the whole length of greenhouse on both sides.

6. **Covering material**

   In early greenhouses, glass has been the main transparent medium used to provide the natural light in protected environments for plant growth.

Nowadays plastics used and it is considered to be more economical, easier to install and they provide good light transmission.

Criteria for Design and Construction of Greenhouses for specific climatic conditions

Depending on the local climate and bioclimatic requirements of species to be cultivated, it is necessary to choose proper site, cladding material, type of structure and architectonic shape of the greenhouse. If the predictable climate generated by the greenhouse is not appropriate complementary facilities and equipment for climate control have to be considered.

i. Greenhouses design criteria for temperate climate

The most limiting climate conditions for greenhouse cultivation in temperate regions are

a) Low night temperatures in winter

b) High daytime temperature

c) High ambient humidity at night and low values during the day; and

d) CO depletion during the day

Therefore, it is necessary to achieve efficient ventilation, which allows for alleviation of the thermal excesses and extreme humidity and prevents CO deficiency. Depending on the type of greenhouse and climate conditions it is advisable that the ventilation area is up to 30% of the ground area of the greenhouse. The increased use of insect-proof screens in the vents to avoid or limit the entrance of insects, decreases the efficiency of ventilation. Collection of rainfall water must also be a priority. In the low-cost type greenhouses, the general problem of condensed water dripping is aggravated in flat-roof greenhouses, inducing serious plant protection problems as it facilitates the development of diseases. Thermal losses must be limited by choosing a suitable cladding material and making it as airtight as possible. Night heating may be necessary for the crop, during the critical winter months but its economic profitability is questionable in many cases.

ii. Greenhouses design criteria for humid tropical climates

The high rainfall during whole year or during rainfall season (which induces high RH), the stability of temperatures (high during both the day and the night) throughout the year and solar radiation (which may be excessive in some cases), are the most outstanding characteristics of humid tropical climates.

In these greenhouses, protection against rainfall must prevail (the greenhouse umbrella effect) and there should be efficient permanent ventilation (with vents frequently equipped with screens to prevent the entrance of insects), as well as a good height and sufficient resistance to withstand strong hurricane winds which are usual in such climates. The designs given in the Fig.1 shows some of the solutions for humid tropical climates. Achieving a compromise between these requirements, at a low cost, is not easy.

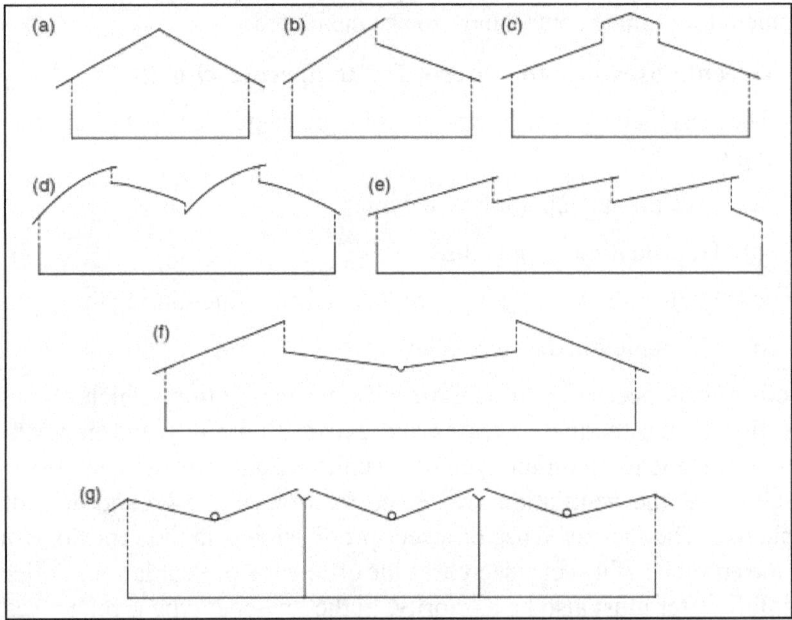

Fig. 1 : Greenhouse structures used in tropical regions

(*Source* : Nicolas Castilla, 2013)

iii. Greenhouses design criteria for other climate conditions

In dry desert climate, the extreme temperature values are more acute than those experienced temperate climates, and the ambient humidity is notably lower and winds being frequently loaded with sand and with very low water content. In these conditions, high ventilation capacity and efficiency is a priority (with the possibility of tightly closing the vents), and there is possibly a need for humidification systems (if the evapotranspiration of water is insufficient) to decrease the temperature and increase the RH. Preventing thermal losses at night is necessary (so choice of a proper cladding material and enough sealing are important) to avoid the need for night heating. The structural resistance to the wind is fundamental and collection of rainfall water for irrigation is normally desirable. Under cold climate conditions, the greenhouse effect must be enhanced and maximum solar energy collection (interception) should be reached with proper roof geometry and cladding material as well as optimized greenhouse orientation. Limiting thermal losses is always desirable (using proper cladding material, thermal screens and being as airtight as possible).

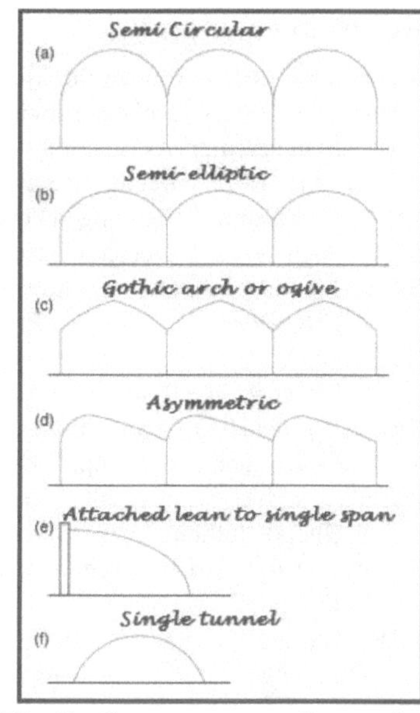

Fig. 2 : Common type of curved roof greenhouses

Frequently, the insulation measures to reduce thermal losses imply a decrease in available solar radiation (the double wall decreases the transmission, the thermal screens generate shadows even when folded) so it is not easy to obtain a compromise solution which must be based on profitability criteria in each specific case. In these cold climates, the obvious choice between multi-span and single-span greenhouses is clearly for the first type. Heating is a must, not just during the winter months, and ventilation is necessary during the season of high radiation. In some cases, greenhouse cladding with a screen (permeable to air and water) aims at achieving a windbreak effect, a shading effect, or plant protection (limiting the access of pests), when the natural thermal conditions are adequate for crop growth and, therefore, a greenhouse effect is not pursued.

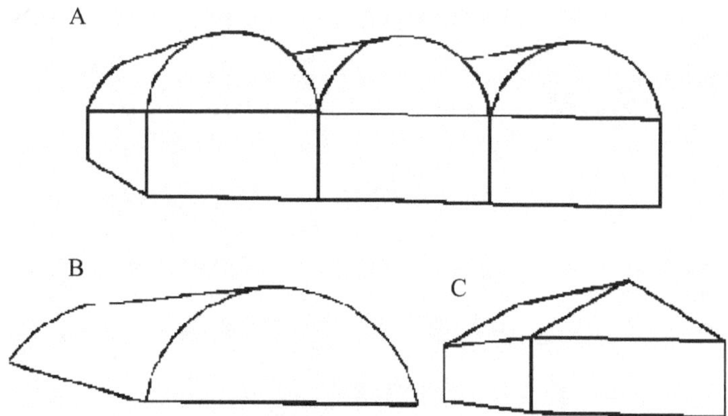

Fig.3 : Commercial greenhouse structures
(a) Gutter connected (b) B' free stancing Quonset (c) Single gable

References

Anonymous. 2015-16. www.indiastat.com

Anonymous. 1994. Greenhouse environment control system considerations. National Greenhouse Manufacturers Association.

Bakker, J.C. 1995. Greenhouse climate control: an Integrated Approach.Wageningen, the Netherlands: Wageningen Pers.

Balraj Singh. 2012. Protected Cultivation of Vegetable Crops. Kalyani Publishers. pp.13-37.

Hanan, J.J. 1998. Greenhouses: Advanced technology for Protected Horticulture. Boca Raton, Fla. CRC Press.

Nelson and V. Paul.1994. Greenhouse operation and Management, Fourth Edition, Prentice Hall, Upper Saddle River, New Jersey, USA.

Nicolas Castilla 2013. Greenhouse Technology and Management. Ediciones Mundi-Prensa, Madrid (Spain) and Mexico.pp.73-75

Prasad, S and U. Kumar. 2005. Greenhouse Management for Horticultural Crops. Agrobios (India), Jothpur.

Sandhya Sharaf. 2012. Greenhouse Management of Horticulture Crops. Mehra Offset Press, Delhi

Vilas, M.S. and A.K. Sharma. 2006. Greenhouse Technology and Applications. S.S. Printers, New Delhi.

4

Artificial Lighting with LEDs in Horticulture

S. Vijaya Kumar and P. Pandiyaraj

Indian Institute of Horticultural Research, Hessaraghatta Lake Post Kandapura, Bengaluru-560089

Abstract

Natural sunlight is the cheapest source available, but for horticulture it is not always attainable in sufficient quantities. Therefore, the use of artificial light has become very common in order to increase production and quality. Under artificial lighting technologies, LED lighting is one of the technology that can help in producing crops and flowers in a more effective and sustainable way around the world. The benefits of LEDs are that the spectrum and intensity can be selected and adjusted, the energy efficiency is higher than most conventional light sources and their small size, durability, long lifetime, cool emitting temperature, and the option to select specific wavelengths for a targeted plant response make LEDs more suitable for plant-based uses than many other light sources. These advantages, coupled with new developments in wavelength availability, light output, and energy conversion efficiency, can bring about a revolution in horticultural lighting. Light quality plays a major role in the appearance and productivity of ornamental and food specialty crop species. Far-red light, for example, is important for stimulating flowering of long-day plants (Deitzer et al., 1979; Downs, 1956) as well as for promoting internode elongation (Morgan and Smith, 1979). Blue light is important for phototropism (Blaauw and Blaauw-Jansen, 1970), for stomatal opening (Schwartz and Zeiger, 1984), and for inhibiting seedling growth on emergence of seedlings from a growth medium (Downs, 1956). The interactions are complex and continue to be unraveled at the molecular level (Devlin et al., 2007), but much of our understanding of these responses comes from studies with narrow-waveband lighting sources, in which LEDs provide obvious advantages. Thus, one potential role of LEDs in horticulture could be to enhance desired characteristics for specific crops. The artificial LED lightings increase the possibility of year-round production of ornamental plants, which will subsequently increase the income of growers related to ornamental industry. Light-emitting diodes pave the way for a better understanding of the interaction between light and pests, diseases, natural enemies and plants, as they make possible the use of very narrow wavelengths. LEDs as an option for reducing the use of growth retardants and obtaining better product quality in ornamental pot plants. Production of secondary metabolites increase with higher blue light ratio. Secondary metabolites production are increased and act against Reactive Oxygen Species (ROS) and pathogens, precondition the plant for environmental changes so they can cope with stress more efficiently.

Keywords: Artificial Lighting, Lighting Sources, Light Emitting Diodes, Energy Efficiency, Ornamental Plants

Introduction

The world floriculture business has reached to almost over a $40 billion mark. It is continuously going thought the rapid transitional changes to have more unique and competitive products of wide choice to the consumers. Netherland is ahead in floriculture trade. Throughout the year, a large number of ornamental commodities are imported in Holland and exported back to the countries for its retail and consumption mostly to the countries in the Gulf, Middle- East Asia and America. According to the statistical estimates worldwide, a large quantity is consumed in US and the other countries. For an example, the US alone consumers 100 lacks of cut flower a day and approximately 4000 lacks every year, though it ranks only 17th in per capita cut flower consumption. The average Swiss spends Rs 4500 a year on flowers, nearly four times than that spent by the average American beside the fact that the consumption of fresh cut flower and foliage plant is on manifold increase every year resulting in the competition worldwide for growing of exportable eco- friendly products. At the same time there is great environmental consciousness towards transforming floriculture with "natural and unspoiled" system of having the exportable produce into a mass and its ease to transport to any distant market. The system relies greatly on the industry with extensive network of breeder, greenhouse managers, field workers, auction houses, sale representatives, shippers, and florists to offer consumers a variety of choices matching their taste. This is the fact that in the recent time, the commercial growers do not rely on the nature to introduce the genetic mutation that will make them millionaires but depends on the art and scientific knowledge to create a novelty in its own. This has resulted in the seriousness and painstaking effort of the growers to produce the plants in new forms and shape to put forward a new plant with extraordinary marketable characteristics every year, e.g. compact chrysanthemum, kalanchoe, bromeliads, orchids etc. are some of the common products with an unusual and eye- catching hue in the European markets. Similarly, the consumers can purchase gerberas with blooms as salad plates and roses with yard- long stems perfectly suited for valentine- day gifts which mainly they get produced under the specifics and artificial environmental created for specifics growth stages of the plant.

LED technology is one of the technologies that can help in producing crops and flowers in a more effective and sustainable way around the world. The benefits of LEDs are clear: you can choose and/or adjust spectrum and intensity, the energy efficiency is higher than most conventional light sources, there is freedom of product design because of the small size of separate LEDs, there is lack of infrared radiation, and they have a long lifetime. With these characteristics they offer loads of (new) opportunities to use them in horticultural applications. Supply & Demand in the greenhouse industry dictate that product quality and delivery schedule be maintained at high market standard for intensively cultivated food and ornamental crops. Product supply and market demand determine wholesale prices that growers can expect to receive for their horticultural products. Even as growers achieve economies of scale, there continues to be increasing pressure on operating margins to compete for economic viability. Specialty (horticultural) crops represent an important sector of the economy generating approximately 50% of total crop production in the USA (USDA, 2005). Any advantage that growers can leverage to reduce production costs while maintaining product quality and schedule integrity is worthy of consideration.

Light is Essential for Plant Growth

Plants have a completely different sensitivity to light colors than humans. With regard to plant growth, light is defined in terms of small particles, also called photons or quantum. The energy content of photons varies, depending on wavelength (light color spectrum). For one optical energy, almost one and a half as many red photons can be produced compared with blue. This means that often red light sources produce more efficient light photons than blue light sources. However the plant has also various sensitivity for various colors of light, and that influences different light-sensitive activities as well. Using the efficient light sources for plants, effective light recipes are important to obtain the optimal results in plant production.

Plant Growth is Spectrum Dependent

Since plant only use a tiny percentage of the full light spectrum and chlorophyll is being green acts as a "Sun-block" to remove 95 per cent of the unwanted and damaging light wavelengths due to higher irradiance at high temperature (Brown, 1995) resulting in poor growth. On the other hand , red and blue (LEDs) lighting have such narrow band widths the plants without green chlorophyll can still survive and will be safe against the harmful light wavelengths. It is further to note that nearly 100 per cent of the light an LEDs grow light emits is completely absorbed by the plants. LEDs can also be used for 24 hours a day without stressing plants since the chlorophyll does not have to do battle with unwanted light –waves as they produce a very small amount of hurt over the plants canopy. If be measure the quantum of LEDs grow lights in Einstein's it mean that how many photons of light strike an area per second which sometimes is just vague in measuring the quantum absorbed in the plants as tiny fraction. However, the ac power factor or amount of light that gets used in plant growth or chlorophyll production as phytochrome response (Folta and Childers, 2008) can be measured as a micromole $m^{-1}s^{-1}$ in the condition when the light are hardly 95 per cent growth efficiency. Also in that case the 95 per cent light is wasted since plants only use very narrow bandwidths of the total light energy.

Lamp Placement to Increase Lighting Efficiency

In addition to light quality, the position of light sources relative to the photosynthetic surfaces of plants has a large effect on crop productivity. Because the radiation energy intercepted by a surface from a point source is related to the inverse square of the distance between them (Bickford and Dunn, 1972), reducing that distance will have a large impact on the incident light level. Compared with scorching hot, high-intensity discharge emitters, cooler LED emitters can be brought much closer to plant tissues. LEDs, therefore, can be operated at much lower energy levels to give the same incident *PPF* at the photosynthetic surface.

Collaboration between Purdue University and the Orbital Technologies Corporation (Madison, WI) has led to the development of a reconfigurable LED lighting array to reduce electrical input for crop lighting. Massa *et al.* (2005a, 2005b) described a lighting array consisting of 16 "light sicles," each of which contains 20 1-inch^2 "light engines" with numerous printed-circuit LEDs. Each square light engine has columns of red and blue LEDs that are independently current-controlled to allow continuous dimming and color blending capability. The lightsicles can be arranged in a separate, vertical, intracanopy configuration whereby a crop stand of planophile plants such as beans or tomatoes can grow up around and surround the vertical light strips. The LED light engines are energized individually from the bottom up to keep pace with the top of the growing crop canopy. Preliminary crop growth studies were performed with cowpea (*Vigna unguiculata* L. Walp. breeding line IT87D-941-1), a dry-bean crop. When compared with stands grown under horizontal, overhead LEDs, either using the same system reconfigured (successive testing) or from a second system that became available later (simultaneous experimentation), intracanopy-grown cowpea produced a greater amount of biomass, converted a higher percentage of light energy into biomass, and had a greater retention rate of inner-canopy leaves. It compares intracanopy versus overhead lighting for cowpea. Lower leaf senescence and abscission resulting from mutual shading from overhead lighting was virtually eliminated in the intracanopy-lighted stands. (Morrow, 2008)

The biomass produced per kilowatt hour of energy consumed was more efficient for cowpea grown with intracanopy lighting than for stands grown with overhead lighting but all other conditions equivalent (Massa *et al.*, 2006). The two growth systems produced comparable evapotranspiration rates (Russell *et al.*, 2006). 'Triton' pepper plants grown with either intracanopy or overhead R + B LED lighting also developed severe occurrence of foliar edema. Although fruit set occurred, the extensive edema on both leaves and flower buds strongly inhibited photosynthetic productivity. The pepper symptoms were not mitigated by using higher percentages of blue light as occurred for cowpea. Preliminary analysis using additional ultraviolet A (365 nm) "black lights" was inconclusive, most likely as a result of the low energy flux from those lamps and unequal distances from the ultraviolet A source to the photosynthetic surfaces within a stand. 'Persimmon' tomato plants grown under the same LED lamps displayed normal growth without edema, indicating that even within solanaceous species, different susceptibilities to this physiological disorder occur. Further investigation of specific light requirements for normal growth and development of different plant species and cultivars will be required as LED lighting systems develop further.

LEDs and Its Utility

A light emitting diodes (LEDs) is a semiconductor diode that emits incoherent narrow- spectrum of light when electrically biased in the forward direction in the form of an electroluminescence. This light source is usually a small area, often with extra optics added to the chip that shapes its radiation pattern and flow over the plant canopy in different colors. The color of the emitted light depends on the composition and condition of the semiconducting material used which can be infrared, visible or near- ultraviolet. Infect, the light emitting diodes are rectifying semiconductor device which converts electrical energy into electromagnetic radiation. The wavelength of the emitted radiation ranges from the near –ultraviolet to the, that is, from about 400 to over 1500 nanometers. (Bourget, 2008)

This light can be used as a regular household light sources and most commercial light emitting diodes, both visible and infrared, are fabricated from type III – V semiconductors. Wherein, these compounds contain elements such as gallium, indium, and aluminum from column- III of periodic table , as well as arsenic, phosphorus and nitrogen from column-v of the periodic table. There are also LEDs products made of II-VI semiconductors, for example Zn, Se, and related compounds. If we combine them together, these semiconductors possess the proper band gap energies to produce radiation at all required wavelengths. Most of these compounds have direct band gaps and therefore, these are efficient in the conversion of electrical energy into radiation.

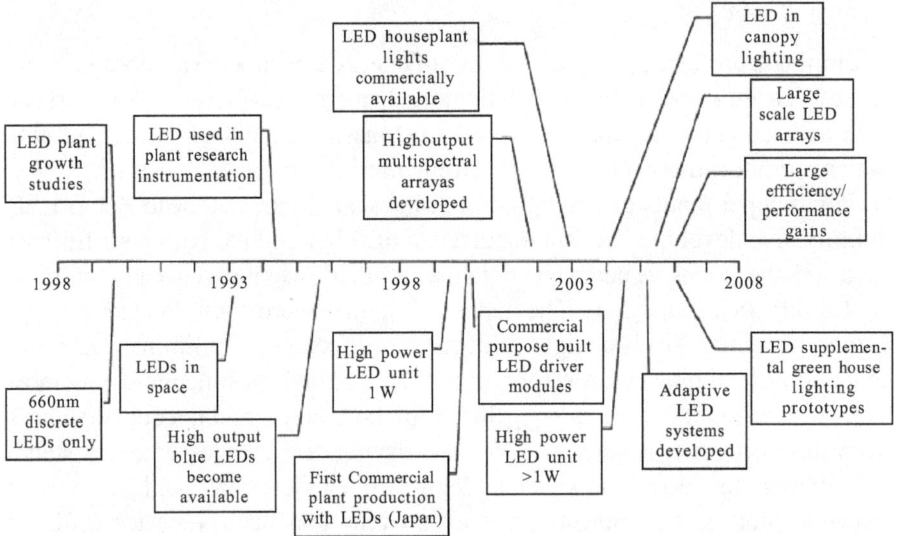

Time Line development of LEDs in horticulture (Morrow, 2008)

The Importance of Blue Light

The use of red LED light to power photosynthesis has been widely accepted for two primary reasons. The red wavelengths (600 to 700 nm) are efficiently absorbed by plant pigments; second, early LEDs were red with the most efficient emitting at 660 nm, close to an absorption peak of chlorophyll. They also saturated phytochrome, creating a high-Pfr photostationary state in the absence of FR or dark reversion. The other main wavelength included in early studies has been in the blue region (400 to 500 nm) of the visible spectrum. The amount of blue light required or optimal for different species is an ongoing question. Blue light has a variety of important photomorphogenic roles in plants, including stomatal control (Schwartz and Zeiger, 1984), which affects water relations and CO_2 exchange, stem elongation (Cosgrove, 1981), and phototropism (Blaauw and Blaauw-Jansen, 1970). The findings showed that wheat could complete a life cycle with red light alone, although added blue light produced larger plants with greater numbers of seeds. One percent blue (H^{-3} μmol·m^{-2}·s^{-1}) was sufficient to keep culm-leaf and flag-leaf length equal to control lengths. However, 10% blue (35 μmol·m^{-2}·s^{-1}) was needed to produce the same number of tillers as control plants (Goins *et al.,* 1997). Shoot dry matter and photosynthetic rates increased with increasing levels of blue light. However, blue light requirements for traits such as stem elongation seem to be genotype-specific, at least in potato (Yorio *et al.,* 1998). Although the potato work was not carried out with LEDs, it has implications for the use of narrow-waveband LEDs in horticultural crop production. It is possible that certain cultivars might grow well with less costly and more efficient single-wavelength LED lighting systems.

Importance of Far-red and Infrared LEDs

The spectral quality of photoperiodic lighting can influence flowering responses. Light quality is perceived by three identified families of plant photoreceptors: cryptochromes, ultraviolet receptors, and phytochromes (Kami *et al.,* 2010). Cryptochromes have been identified in many plant species and mediate a variety of light responses, including playing a role in flowering time regulation in Arabidopsis thaliana (Cashmore *et al.,*1999; Mockler *et al.*, 2003). In photoperiodic crops, the PFR/PR+FR, through different types of phytochromes, influences flowering. PFR is the active form of phytochrome, which translocates to the nucleus on receiving light signals and activates downstream pathways (Franklin and Quail, 2010). Under a long, uninterrupted night, the PFR form of phytochrome slowly converts to the PR form, leaving insufficient PFR to inhibit flowering. However, if R light is provided during the long night, PR is converted to PFR (creating a greater PFR/PR+FR), which inhibits flowering in SDPs. Several other studies using FR LEDs are discussed by Kim *et al.* (2005) examining plant morphology, disease development, and nitrate accumulation. Until recently, it was difficult to obtain LED arrays with a wide light spectrum tunable at different emission peaks, but with the rapid, ongoing development of LED technology, such studies can now be conducted using multispectral arrays that generate a variety of colors or even white light.

Importance of LEDs in Horticulture

LEDs can play a variety of roles in horticultural lighting. They are particularly well suited for research applications (e.g., in growth chambers) as a result of their unique capabilities and lend themselves well to shelf lighting for tissue culture applications as a result of their low profile and low radiant heat output. LEDs can also be used for production lighting in controlled environments and supplemental and photoperiod lighting in greenhouses, although these applications are in the early stages of development. The ability to control spectral quality is also of interest for plant display lighting (Morrow, 2008). Light quality plays a major role in the appearance and productivity of ornamental and food specialty crop species. Far-red light, for example, is important for stimulating flowering of long-day plants (Deitzer *et al.*, 1979; Downs, 1956) as well as for promoting internode elongation (Morgan and Smith, 1979). Blue light is important for phototropism (Blaauw and Blaauw-Jansen, 1970), for stomatal opening (Schwartz and Zeiger, 1984), and for inhibiting seedling growth on emergence of seedlings from a growth medium (Thomas and Dickinson, 1979). The blue light photoreceptor class of cryptochromes has been found to work in conjunction with the red/FR phytochrome photoreceptor class to control factors such as circadian rhythms and de-etiolation in plants (Devlin *et al.*, 2007). The interactions are complex and continue to be unraveled at the molecular level (Devlin *et al.*, 2007), but much of our understanding of these responses comes from studies with narrow-waveband lighting sources, in which LEDs provide obvious advantages. Similar studies have been performed with materials that modify intercepted light quality such as colored films, mulches, or Color Nets, crop netting (Polysack Plastics Industries, Israel) (Shahak *et al.*, 2004). Thus, one potential role of LEDs in horticulture could be to enhance desired characteristics for specific crops.

In addition to changes in appearance and productivity, plant responses to narrow-bandwidth light sources or to supplemental LED lighting range from decreased viral resistance in pepper to increased suppression of pathogens in tomato and cucumber to increased nitrate accumulation in spinach (Kim et al., 2005 and references therein). These studies are just the tip of an as yet unmapped iceberg of crop responses to narrow-spectrum lighting. Future needs for controlled environment crop management also will involve interactions of lighting parameters with still other environmental factors. Crop breeders could, for example, select phenotypes with desirable traits expressed in response to unique lighting conditions. What are some possible advantages of tailored light quality and application methods? From the data currently available, it seems likely that custom-designed lighting systems could significantly reduce insect, disease, or pathogen loads on certain crops. It is easy to imagine a lighting system enhanced or restricted in certain wavelengths that eliminates or minimizes the abilities of fungi to proliferate or insects to navigate to host species, reproduce, and so on. Although these advantages might be limited in a commercial production setting, they could be significant in growth facilities for disease-free germplasm production. Other easy-to-imagine scenarios include using select-waveband LEDs to stimulate early or uniform flowering in seasonal ornamentals or to generate specialty produce crops with enhanced levels of vitamins or minerals. Possibly treating crops with low dosages of narrow-band radiation at key points in the life cycle could initiate a cascade of responses in a cost-effective manner. Indeed, as LED technology continues to develop and the price of components drops, LEDs may fill many, if not all, niches of other more traditional horticultural light sources. Anticipating such eventualities now will allow horticulturists to keep pace with advances in this rapidly developing technology.

LEDs for Sustain Plant Growth

LED lights are increasingly becoming widespread in terms of additional sources in plant cultivation. Advantages make LEDs perfect for supporting plant growth in controlled environment such as greenhouse cultivation, plant tissue culture room and growth chamber. Depending on its own features, each plant species reacts differently to the light source and light wavelength (Koksal, 2015). Plant growth under red LEDs on rose were reported by Bula *et al.* (1991). Martineau *et al.* calculated that the amounts of dry matter per mole of artificial lighting gained by rose grown using red (650 nm) LEDs or high-pressure sodium lamps were identical, and Chang *et al.* calculated that the maximum photon utilization efficiency for growth of the green alga *Chlamydomonas reinhardtii* under red LEDs is centred at 674 nm. Lettuce grown under red LEDs presented hypocotyls and cotyledons that were elongated, a phenomenon known to be phytochrome-dependent. Under red LEDs illumination, phytochrome stimulation is especially high as far red light is not provided. Hypocotyl elongation could be prevented by adding at least 15 $\mu mol\ m^{-2}\ s^{-1}$ of blue light. Although a complete demonstration was not provided, one can hypothesize that the supplemented blue light activated cryptochrome, a blue-light photoreceptor that mediates reduction of hypocotyl length. The efficiency of red (650–665 nm) LEDs on plant growth is easy to understand because these wavelengths perfectly fit with the absorption peak of chlorophylls and phytochrome, while the supplemented blue light introduced the idea that growth under natural light could be mimicked using blue and red LEDs. In addition to providing a better excitation of the different types of photoreceptors, the blue + red combination allowed a higher photosynthetic activity than that under either monochromatic light. Some authors attributed this effect to a higher nitrogen content of the blue-light-supplemented plants, whereas others suggested a better stomatal opening, thus providing more CO_2 for photosynthesis. It is well established that stomata opening is controlled by blue-light photoreceptors.

Commercial uses of LEDs in Landscape Horticulture

LED light can be successfully incorporated into a landscaping in variety of ways. Specifically, LED strips are effective for lighting long paths, stairs or pools, tank to their lengthy design. Throw down a long strip, and a large amount of real estate can be covered and properly lit. Mounted LED fixture that is strategically placed can illuminate a specific spot a lawn. LED can also be installed directly into other devices, such as lanterns, for enhanced lighting, also, thanks to their immense flexibility; LEDs can be brighter or softer depending on your security needs or the mood that needs to be set. Voltages and lamps for the LEDs are highly configurable. In term of a landscape job, this means that non-intrusive, low- light bulbs can be used for illuminating footpaths and sidewalks, while brighter light can be hours just outside the home , or on a stoop. Additionally, lights of widely different colors can be used to spice up a landscape lighting even a nighttime lighting is becoming common in and outdoor, almost at no cost is elegance beyond imagination. The following are kind of styles recently being used in gardening.

LEDs garden stick lights are being used mainly for energy saving, long working life and easy "stick and play" installation along weather – proof and super bright LEDs light properties. These lights serve as great supplement for garden or home decoration these LEDs garden sticks lights. Can be solar powered as super bright LED color changing light having almost no electric cord wiring, long working hours in an aluminum stick body and durable weather – proof super bright LEDs light bulbs. These light normally are build in two solar panels and with one AAA Ni- Nh rechargeable decoration and perfect gifts for Christmas and holyday seasons. The most valuable advantage of this kind kid of lights in the garden is that they can be put simply in water as plug and play, easy to relocate and need no hardware installation. These make a great ease for display in pond , party event or it can be sit on the regular fountain to make a beautiful lighted water spray. The solar power LED lights are large variants of solar garden light e.g., solar fairy lights with changing colors and having almost 100 LED bulbs. These bulbs are also fixed as solar powered illuminated rock lights to provide an ambient lighting for the garden or patio. These garden lights save energy while the lighting unit illuminates automatically. In general, solar garden lights, as with many things, new solar LEDs technology allows light discharge of up to 12 hours per night. The individual advantage is that these solar energy LED garden lights require no wiring as they charge through the daytime from the sun's rays (they require a sunny position, although will gain some charge on an overcast day). The sola based LED panels are also highly efficient , mono – crystal silicon solar based panel and have rechargeable battery (Ni- Cd /Ni-MH), preferable and adjustable LED color e.g., super bright white, yellow, green, red, blue etc. (Massa *et al.,* 2006) and the all lights in panel are in automatically turned on mode. The new fancy LEDs light have been developed in a matching shape of safe objects to give an impression of sea – life by 7 illumination objects in open garden corners. These lights look like tiny sea urchin shells and often are available at beach shops in US and Europe. These water garden landscape lights are thin and lighten as eggshells. While using solar power based LED lights in the garden ultra –low cost temporary garden lights using LEDs, lithium coin cells, and mason jars can be made to start with for which we need LEDs for this design is an ultra-bright LED with a diffused lens available in commercial stores and they are effective in casting the light by LEDs chip every direction. One may also opt for an easy access, with 10mm diffused white LEDs and there are still similar LEDs available elsewhere as well.

Commercial use of LEDs in Plant Propagation

Among all the light intensities intercepted by the plants red and blue light is essential for plants growth as it is a fact that the leaves perform the process of photosynthesis and red and blue spectrum or wavelengths are most essential for photosynthesis. The blue light is absorbed to its maximum in the spectrum region of 430-450 nm while the red in the spectrum with 650-670 nm. It has been estimated that the ideal balance for photosynthesis is 92 per cent red LEDs and 8 per cent blue LEDs if it is practiced artificially. In this case the blue lights have a smaller influence than those of the red lights and it makes easy to select anything between 1 per cent and 20 per cent of the blue lights, depending on requirements of the plants for their growth and developments. The commercial LEDs light are available as per color and the frequency of light used by plants for efficiency LED lights are available as per color and the frequency of light used by commercial LEDs lights are available as per color and the frequency of light used by plants for efficient and healthy growth so as to provide an efficient photosynthesis energy than the use of conventional grow lights. In general , red light being in the vicinity of a plant in the first peak of light absorption spectrum i.e., 660 nm it contributes maximum to the plants photosynthesis. However, the red light, when combined with the blue light, encourages flowering and most often used in higher doses to stimulate flowering, seeding and fruiting in certain plants of economic importance. Whereas, the blue light is responsible primarily for vegetative growth (leaf) as in fact the chlorophyll has its second distinct absorption peak in the vicinity of the blue light wavelengths (450nm). The blue light is indispensable to the morphology healthy growth of plants.

LEDs Application in Greenhouse Horticulture

In majority of the flowering growing countries, conditions are perfect and the days are always 12 hours long; sunlight is intense, and nights are cool. However, the most commercial sectors are now moving to the still better climate conditions, into the developing countries. The flowers are now the third most important product, after oil and banana in gulf countries and despite the hard economics the flower business is thriving well and there are several flowering plants those response to the natural short day (light) conditions. To cut short their vegetative phase into reproductive with sufficiently elongated stems and quality flower buds we may have provision of providing the plants an extra photoperiodic as well as photosynthetic light to make them accessible to enough exposure, especially in winter. It is evident that every plant contains a photosensors that respond to the specific part of the spectrum from the ultra violet radiation (UV) to 740 nm spectrum which is known to control various aspects to plant growth and flowering. In last decade, several researcher worldwide have studied the effect of light on the plants to explore that which part of the daylight provide most suitable spectrum of photosynthesis. Its further more critical that how low intensities of LEDs lights (on the order of 1 W/m2 or 100µmoles m-2sec -1 in total) can be used to control the desired characteristics in the plants for final uniform compactness and architecture of the plant as whole. It has been already standardized the plants use light from 400 to 700nm for photosynthesis, with some wavelength having the greatest effect at certain times of the day and theirs at various stages of plants growth. The spectrum can affect properties such as compactness of shape and time to flowering and final yields. Not only that in such plants for food and medical use, it can change the plants chemical content, can which affect flavour, color, and nutritional and / or pharmaceutical value.

It is very important to provide to optimize the lighting requirement for a given effect on the plants. For example, the various scientists in Europe and US have revealed that a particular red wavelength causes a chrysanthemum plant to flower four weeks earlier (Downs, 1956) than it does under the same conditions without the additional LED lighting using daylight. It has also be reported that the system could reduce the cost to an economically viable level and is affordable with present – day LEDs. The application of LEDs is commercially highly efficient with available narrow emission spectra across the relevant wavelength range over a plant canopy resulting in an flower bud induction without disturbing the dark period in photo-sensitive plants while using smart LEDs in a mixture of 80 per cent red and 20 per cent blue (Yanangi *et al.,* 1996). Recently introduced method for commercial greenhouse is use for plastics sheeting with special reflectance spectra and placed as reflective mulch on the floor or on benches where pot plants are grown as it s an obvious that at white plastics mulch can be increase light intensity at the plant while absorbing UV. On the other hand, the red plastic mulch is often used to improve the flavour in soft fruit. Whereas the silver color mulch have been found to reflect throughout the near-UV, visible and near – IR lights, are commonly being used on large scale in the greenhouse to improve grape quality and increased productivity.

Role of LEDs in Disease Management

The influence of light can be termed as day length, intensity, quality and integral. Photosynthesis is the major process that makes interaction between plants and light. Photoreceptor is a plant part that helps to absorb light energy from sun. Light energy captured by photoreceptor is a part biological process. Supplementary lighting is very common in northern hemisphere because of short period of day length. Supplementary lighting has been used to grow plants nearly 150 years. Supplementary lighting can suppress the growth of microbial pathogens by which we can control the diseases of plants (Shahak, 2004). Plants were grown under light-emitting diode (LED) arrays with various spectra to determine the effects of light quality on the development of diseases caused by tomato mosaic virus (ToMV) on pepper (*Capsicum annuum* L.), powdery mildew [*Sphaerotheca fuliginea* (Schlectend:Fr.) Pollaci] on cucumber (*Cucumis sativus* L.), and bacterial wilt (Pseudomonas solanacearum Smith) on tomato (*Lycopersicon esculentum* Mill.). One LED (660) array supplied 99% red light at 660 nm (25 nm bandwidth at half-peak height) and 1% far-red light between 700 to 800 nm. A second LED (660/735) array supplied 83% red light at 660 nm and 17% far-red light at 735 nm (25 nm bandwidth at half-peak height). A third LED (660/BF) array supplied 98% red light at 660 nm, 1% blue light (BF) between 350 to 550 nm, and 1% far-red light between 700 to 800 nm. Control plants were grown under broad-spectrum metal halide (MH) lamps. Plants were grown at a mean photon flux (300 to 800 nm) of 330 micromoles m-2 s-1 under a 12-h day/night photoperiod. Spectral quality affected each pathosystem differently. In the ToMV/pepper pathosystem, disease symptoms developed slower and were less severe in plants grown under light sources that contained blue and UV-A wavelengths (MH and 660/BF treatments) compared to plants grown under light sources that lacked blue and UV-A wavelengths (660 and 660/735 LED arrays). In contrast, the number of colonies per leaf was highest and the mean colony diameters of S. fuliginea on cucumber plants were largest on leaves grown under the MH lamp (highest amount of blue and UV-A light) and least on leaves grown under the 660 LED array (no blue or UV-A light). The addition of far-red irradiation to the primary light source in the 660/735 LED array increased the colony counts per leaf in the *S.fuliginea/* cucumber pathosystem compared to the red-only (660) LED array. In the P. solanacearum/tomato pathosystem, disease symptoms were less severe in plants grown under the 660 LED array, but the effects of spectral quality on disease development when other wavelengths were included in the light source (MH-, 660/BF-, and 660/735-grown plants) were equivocal. These results demonstrate that spectral quality may be useful as a component of an integrated pest management program for future space-based controlled ecological life support systems. (Shahak *et al.,* 2004)

LEDs in Insect Management

Most insects have two types of photoreceptive organs, compound eyes and ocelli. Compound eyes are made up of a large number of light-sensitive units termed ommatidia. An ommatidium contains an elongated bundle of photoreceptor cells, each having specific spectral sensitivities. The ommatidia are packed in a hexagonal array so as to cover a large visual field with certain spatial resolution and to perceive the motion of objects. The spectral sensitivities of photoreceptors determine the visible light wavelength for insects, which often expands into the colored region, which is invisible to humans. A compound eye typically contains three types of photoreceptor cells with spectral sensitivity peaking in the red, blue, and green wavelength regions, respectively, as exemplified in honeybees (Post and Gold Smith, 1965). Fruit-piercing moths such as *Eudocima tyrannus* Guenée and *Oraesia emarginata* Fabricius damage fruit in orchards. Damage can be prevented by running yellow fluorescent lamps in the orchard at night (Nomura 1967; Nomura *et al.,* 1965). This strategy makes use of the fact that when moths encounter light above a certain brightness at night, under which their compound eyes are light-adapted as in the daytime (Meyer-Rochow 1974; Walcott 1969), the light adaptation suppresses nocturnal behaviors such as flying, sucking the juice of fruit, and mating. This technique of suppressing behavior using yellow fluorescent light is also used to prevent damage to chrysanthemums and carnations by the cotton bollworm *Helicoverpa armigera* Hübner (Yase *et al*., 1997), damage to green perilla by the common cutworm *Spodoptera litura* (Fabricius), and damage to cabbage by the webworm, *Hellula undalis* Fabricius (Yase *et al.,* 2004). The light-adapted and dark-adapted states of the compound eye of adult *Helicoverpa armigera*. Recently, green fluorescent lamps have also been developed for the control of nocturnal moths. These lamps suppress the behavioral activity of a number of moth species in the same way as yellow fluorescent lamps but have little effect on the growth of plants compared to the yellow lamps (Yamada *et al.,* 2006; Kono and Yase 1996; Yase *et al*., 1997). Furthermore, because LED lighting is becoming considerably cheaper, yellow-emitting LEDs have been recently applied to control the behaviors of nocturnal moths (Hirama *et al.,* 2007; Yabu 1999; Yoon *et al.,* 2012). LEDs can produce highly monochromatic lights (i.e., with a narrow range of wavelength) across the spectrum from UV to red. This optical characteristic of LEDs is an advantage for controlling pest behavior and their practical application is expected in the near future.

Role of LED in Tissue Culture

In-vitro culture method has been practiced as a technique to various types of ornamental plant especially for flower colour, plant morphology and some physiological characters. Primary phases for *in vitro* involve sterilization, multiplication and hardening. Moreover, it is important to produce disease free explant with rapid multiplication rate to ensure successful cultivation of particular crop. According to Preil *et al.* (1985) *in vitro* culture probably shorten breeding cycles and decrease the development cost. Light source such as fluorescent lamps, metal halide lamps and high-pressure sodium lamps are generally used for in vitro cultures. However, these lights contains unnecessary wavelength that are low in quality for stimulating growth. Furthermore, the use of these lights consumed a lot of electricity and it was reported that tissue cultured lab consumed 65% of total electricity. Light Emitting Diode also known as LED light has been utilized globally in agriculture as an alternative light source for plant growth and photosynthesis. LEDs have attracted considerable interest because of their wavelength specificity and narrow bandwidth, small mass and volume, long-life and minimum heating.

Morphology and physiology of in vitro grown plants are regulated by various micro-environmental factors such as light, temperature, humidity and carbon dioxide. Inadequate or insufficient light may cause harmful for plant growth or led to excessive growth respectively. Light quality has effect on morphological characteristics such as stem elongation, leaf size, plant anatomy that shows the important of light as a promising source The previous research had successfully showed the significance of red (R) and blue (B) LED lights on plant photosynthesis. R and B color of LED has been widely used to enhance plant growth. However, there were few studies involved in vitro plant and rose as a subject. Past studies had done in vitro technique under RB ratio 1:1. Xue Fan *et al,* reported to apply RB with ratio 1:1 for leaf development of young tomato and Li *et al.*, applied RB with ratio 1:1 with wavelength 660nm and 460nm respectively on rapeseed. Different spectrum of light truly gave an impact in term of anatomy, physiology and morphology of leaves, Nevertheless, the spectrums emitted from both colours vary too much with the sunlight. Within the range 400-700nm, light spectrum is able to provide the most important energy for plant photosynthesis among the solar energy. Studies by Matsuda et al., rice plants that grown under combination of blue (470nm) and red (660nm) shows greater leaf photosynthetic rates compared to single colour LED. A study by Harun et al., reported that, treatment under 16:4 RB ratio, it is more effective to promote higher photosynthesis activity which lead to increase number of shoots and leaf. High photon level of blue and red LED and their wavelength have certainly given an advantage. Therefore, the objective of this study is to examine the effect of red blue LED with ratio 16:4 in multiplication of *Rosa kordesii* in tissue culture as suggested by Tanaka, *et al.,* that LED has improved the quality of explant to increase shoot multiplication.

Role of LEDs in Secondary Metabolites Production

Photosynthetic organisms exposed to high light develop short and long term response mechanisms to reduce stress effects. Some of these mechanisms are the specific topic of other papers included in this special issue (xanthophyll cycle, non-photochemical quenching, re-oxidation of the reduction equivalents through photorespiration, the malate valve and the action of antioxidants). This section is dedicated to the metabolic shifts triggered by high light stress. They are used in repairing mechanisms, shielding reactive oxygen species (ROS) quenching or the production of storage compounds (Bickford and Dunn, 1972). The synthesis of the metabolites takes place in plastids (terpenoids) or involves them (phenylpropanoids. Typical examples are medicinal plants and herbs of pharmaceutical importance such as mint (*Mentha* sp.) and jewel orchid (*Anoectohilus* sp.) However, a decrease in secondary metabolites, flavonoids and phenolics, was also observed with increasing irradiance in the medicinal plant cat's whiskers (*Orthosiphon stamineus*) indicating that the light irradiance may have negative consequences on secondary metabolite production. In higher plants, it has been documented that depending on species and growing conditions, the secondary metabolites and pigments in the flavonoid family accumulate under photoinhibitory conditions at cell level, although the mechanistic aspects of LED light effects are not well understood.

The high influence effect of LED light has been studied more in photosynthetic microorganisms, partly because they present huge biotechnological and economic potential (biofuels, pharmaceuticals, food additives and cosmetics). Additional red or blue (470 nm) LED light caused stress whereby the xanthophyll cycle was activated. The additional blue light was less stressful than the red light. More recently, Barta *et al.* (1992) showed that in mixotrophic growing conditions, flashing LED light (8 μmol photon m^{-2} s^{-1}) triggered similar astaxanthin concentration to continuous LED light (12 μmol photon m^{-2} s^{-1}). Such low light requirement suggests the involvement of photoreceptors. A putative transduction mechanism of the blue light signal would involve major carotenoids in *D. salina*. Signalling of secondary carotenoid synthesis involves chloroplast-generated ROS. Much more investigation is needed to understand the impact of LED light on primary and secondary metabolism of photosynthetic organisms.

Conclusion

Artificial lightings increases the possibility of year-round production of ornamental plants. Uses of solid-state lighting over other lamp technology include the ability to provide high light intensities with low radiant heat, adjustable spectral quality that allows optimization to improve photosynthetic efficiency and plant form and function, good safety characteristics, and operating capabilities that can significantly reduce power use. LED lightning system can provide precision delivery of photons and more cost effective. Light-emitting diodes pave the way for a better understanding of the interaction between light and pests, diseases, natural enemies and plants, as they make possible the use of very narrow wavelengths. Use of growth retardants and obtaining better product quality in ornamental pot plants by LEDs. Production of secondary metabolites increase with higher blue light ratio.

References

Barta, D.J., Tibbitts, T.W., Bula, R.J., Morrow, R.C. (1992) Evaluation of light emitting diode characteristics for space-based plant irradiation source. Adv. Space Res. 12:141–149.

Bickford, E.D., Dunn, S. (1972) Lighting for plant growth (The Kent State Univ. Press, Kent, OH).

Blaauw, O., Blaauw-Jansen, G. (1970) The phototropic responses of Avena coleoptiles. ActaBotan.Neer. 19:755–763.

Bourget, C.M. (2008) An introduction to light-emitting diodes. Hort Science 43:1944–1946.

Brown, C.S., Schuerger, A.C., Sager, J.C. (1995) Growth and photomorphogenesis of pepper plants grown under red light-emitting diodes supplemented with blue or far-red illumination. J. Amer. Soc. Hort. Sci. 120:808–813.

Bula, R.J., Morrow, R.C., Tibbitts, T.W., Barta, D.J., Ignatius, R.W., Martin, T.S. (1991) Light-emitting diodes as a radiation source for plants. HortScience 26:203–205.

Cosgrove, D.J. (1981) Rapid suppression of growth by blue light. Plant Physiol. 67:584–590.

Deitzer, G.F., Hayes, R., Jabben, M. (1979) Kinetics and time dependence of the effect of far red light on the photoperiodic induction of flowering in Wintex barley. Plant Physiol. 64:1015–1021. Abstract/FREE Full Text

Devlin, P.F., Christie, J.M., Terry, M.J. (2007) Many hands make light work. J. Expt. Bot. 58:3071–3077.

Downs, R.J. (1956) Photoreversibility of flower initiation. Plant Physiol. 31:279–284.

Folta, K.M., Childers, K.S. (2008) Light as a growth regulator: Controlling plant biology with narrow-bandwidth solid-state lighting systems. HortScience 43:1957–1964.

Goins, G.D., Yorio, N.C., Sanwo, M.M., Brown, C.S. (1997) Photomorphogenesis, photosynthesis, and seed yield of wheat plants grown under red light-emitting diodes (LEDs) with and without supplemental blue lighting. J. Expt. Bot. 48:1407–1413.

Hirama, J., Seki, K., Hosodani, N., Matsui, Y. (2007) Development of a physical control device for insect pests using a yellow LED light source: results of behavioral observations of the Noctuidae family. J SHITA 19:34–40.

Kim, H.H., Wheeler, R.M., Sager, J.C., Yorio, N.C., Goins, G.D. (2005) Light-emitting diodes as an illumination source for plants: A review of research at Kennedy Space Center. Habitation (Elmsford) 10:71–78.

Koksal, N., Incesu, M and Teke, A. 2015. Supplemental LED lighting increases pansy growth. Horticultura Brasileira 33: 428-433.

Kono, S., Yase, J. (1996) Characteristic of physical control and using technology. Utilization of color sense of insects. Prant Prot. 50:30–33.

Massa, G.D., Emmerich, J.C., Mick, M.E., Kennedy, R.J., Morrow, R.C., Mitchell, C.A. (2005a) Development and testing of an efficient LED intracanopy lighting design for minimizing equivalent system mass in an advanced life-support system. Gravit. Space Biol. Bull. 18:87–88.

Massa, G.D., Emmerich, J.C., Morrow, R.C., Mitchell, C.A. (2005b) Development of a reconfigurable LED plant-growth lighting system for equivalent system mass reduction in an ALS (SAE Technical Paper 2005-01-2955).

Massa, G.D., Emmerich, J.C., Morrow, R.C., Bourget, C.M., Mitchell, C.A. (2006) Plant-growth lighting for space life support: A review. Gravit.Space. Biol. 19:19–29.

Meyer-Rochow, V.B. (1974) Fine structural changes in dark-light adaptation in relation to unit studies of an insect compound eye with a crustacean-like rhabdom. J Insect Physiol 20:573–589

Morgan, D.C., Smith, H. (1979) A systematic relationship between phytochrome-controlled development and species habitat, for plants grown in simulated natural irradiation.Planta 145:253–258.

Morrow, R.C. (2008) LED lighting in horticulture.Hort Science 43:1947–1950.

Morrow, R.C., Duffie, N.A., Tibbitts, T.W., Bula, R.J., Barta, D.J., Ming, D.W., Wheeler, R.M., Nomura K (1967) Studies on orchard illumination against fruitpiercing moths. III. Inhibition of moths' flying to orchard by illumination. Jpn J Appl Entomol Zool 11:21–28 (in Japanese with English summary)

Nomura K, Oya S, Watanabe I, Kawamura H (1965) Studies on orchard illumination against fruit-piercing moths. I. Analysis of illumination effects, and influence of light elements on moths' activities. Jpn J Appl Entomol Zool 9:179–186 (in Japanese with English summary)

Post CT, Goldsmith TH (1965) Pigment migration and lightadaptation in the eye of the moth, Galleria mellonella. Biol Bull 128:473–487

W. Preil, W. Horn, C,J. Jensen, W, Odenbach, and O. Scheider. 1985. In vitro propagation and breeding of ornamental plants: advantages and disadvantages of variability. In: (eds) Genetic Manipulation in Plant Breeding.1986, (pp 377–403) Proc. EUCARPA Int. Symp. Berlin, EUCARPA, Berlin

Russell, J.F.. Massa, G.D.. Mitchell, C.A.. Sager, J.C., McFarlane, J.C. (1997) in Plant growth chamber handbook, Radiation, edsLanghans R.W., Tibbitts T.W. (Iowa State Univ. Press: North Central Region Research Publication No. 340, Iowa Agriculture and Home Economics Experiment Station Special Report no. 99, Ames, IA), pp 1–29.

Schuerger, A.C., Brown, C.S., Stryjewski, E.C. (1997) Anatomical features of pepper plants (Capsicum annuum L.) grown under red light-emitting diodes supplemented with blue or far-red light. Ann. Bot. (Lond.) 79:273–282.

Schwartz, A., Zeiger, E. (1984) Metabolic energy for stomatal opening: Roles of photophosphorylation and oxidative phosphorylation. Planta 161:129–136.

Shahak, Y., Gussakovsky, E.E., Gal, E., Ganelevin, R. (2004) ColorNets: Crop protection and light-quality manipulation in one technology. Acta Hort. 659:143–151.

Thomas, B., Dickinson, H.G. (1979) Evidence for two photoreceptors controlling growth in de-etiolated seedlings.Planta 146:545–550.

Walcott, B. (1969) Movement of retinula cells in insect eyes on light adaptation. Nature 223:971–972.

Yabu, T. (1999) Control of insect pests by using illuminator of ultra high luminance light emitting diode (LED). Effect of the illumination on the flight and mating behavior of Helicoverpa armigera. Plant Prot 53:209–211.

Yamada, M., Uchida, T., Kuramitsu, O., Kosaka, S., Nishimura, T., Arikawa, K. (2006) Insect control lighting for reduced and insecticide-free agriculture. Matsushita-Denko-Giho 54(1): 30–35 (in Japanese with English summary)

Yanagi, T., Okamoto, K., Takita, S. (1996a) Effect of blue and red light intensity on photosynthetic rate of strawberry leaves. Acta Hort. 440:371–376.

Yase, J., Yamanaka, M., Fujii, H., Kosaka, S. (1997) Control of tobacco budworm, Helicoverpa armigera (Hubner), beet armyworm, Spodoptera exigua (Hubner), common cutworm, Spodoptera litura (Fabricius), feeding on carnation, roses and chrysanthemum by overnight illumination with yellow fluoresent lamps.Bull Natl Agric Res Cent West Reg 93:10–14 (in Japanese)

Yase J, Nagaoka O, Futai K, Izumida T, Kosaka S (2004) Control of cabbage webworm, Hellula undalis Fabricius (Lepidoptera: Pyralidae) using yellow fluorescent lamps. Jpn J Appl Entomol Zool Chugoku Bra 46:29–37 (in Japanese with English abstract)

Yoon, J.B., Nomura, M., Ishikura, S. (2012) Analysis of the flight activity of the cotton bollworm Helicoverpa armigera (Hubner)(Lepidoptera: Noctuidae) under yellow LED lighting. Jpn J Appl Entomol Zool 56:103–110.

Yorio, N.C., Wheeler, R.M., Goins, G.D., Sanwo-Lewandowski, M.M., Mackowiak, C.L., Brown, C.S., Sager, J.C., Stutte, G.W. (1998) Blue light requirements for crop plants used in bioregenerative life support systems. Life Support Biosph. Sci. 5:119–128.

5

Technology for Gerbera Flower (*Gerbera Jamesonii*) Cultivation UnderNatural Ventilated Polyhouse

T. K. Chowdhuri, R. Sadhukhan and S. Das

Associate Professor, Department of Floriculture and Landscaping
Professor, Department of Genetics and Plant Breeding
Assistant Professor, Department of Agronomy, Bidhan Chandra Krishi Viswavidyalaya Mohanpur, Nadia, West Bengal - 741252

Introduction

Gerbera is one of the most important commercial flowering crops growing in the world as a cut flower as well as garden decoration, which belongs to family compositae and native in South Africa and Asiatic region. Presently, its demand and utility are rising up continuously in India both in urban as well as rural areas throughout the year. Peoples prefer this flower, because of its unparallel diversity in colour and long vase life; whereas farmers took this crop in relation to high margin of profit and long post harvest life over some others flower. Now, people are showing their keen interest towards use of this flower in their daily life and for celebrating a colorful festival. As per report of NHB (2014), total crop area is 820 hactors, where flower marketed of 3.96 and 17.84 MT of loose and cut flowers respectively. The prime producing states in India for gerbera flower production are Karnataka, Maharastra, Assam, Uttrakhand, Telangana, Tamilnadu and West Bengal. There are some constrains identified in gerbera cultivation under protected situation, like initial investment is too much, protocol of the technology is very much restricted, regular supervision is an essential and all local festivals regulates quantity of consumption etc. However, before adopting this technology for implementation as taking part of business, users should upgrade their knowledge by seeing and believing at different production centers. Details about latest technology are summarized under mentioned, which may be helpful to the students and farmers to some extent for ready references.

Species and Cultivars

The genus of Gerbera consists of about forty species of half hardy, perennial flowering plants reported by Das *et al* (1999). Besides the cultivars growing in garden decoration, there are thousands of commercial cultivars developed by the National Institutes, SAUs of AICRO on Floriculture and private sectors with attractive colours. There are selected various potential varieties of gerbera grown under natural ventilated poly house in different parts of India like **White colour**: Balance, Donna Tella,

Yellow colour: Paradiso, Brillance, Dana Ellen, Fredeking, Nadja, Uranus,

Orange colour : Galiath, Infarno, Maron Clementine, Nitta, Sunburst, Dimond Jubilee,

Fla-Orange, Avo-Gino, Hawaii, Mauritius Orange, Horning Orange, Horning Rubin etc.

Pink colour: Abe Pink, Blush, Candy Strip, Rosalin, Pree Intenzz, Pink Elegance, Flemingo, Fredaisy, Terraqueen, Valentine etc.

Red: Ozaki, Kozohara, Kaumana, Toyamo, Pohoa, Hayashii, Asahi, Mickey Mouse, CanCan, Avo-Rosette, Jacqueline etc.

Brick Red colour: Walhalla and Red colour: Jaffana, Stranza, Dusty, Fredorella, Vesta.

Growth and Development

It is very much an essential to know about growth and development of gerbera by the technologist or a farmer for running a commercialization unit. Gerbera plants are propagated commonly by separation of suckers, but for obtaining maximum yield, tissue culture plant materials are the best. Planting of gerbera is an ideal before or after rainy season in polyhouse. After planting of three months, when plants were having 15-16 nos. of leaves are able to produce flower. It has been found that flower production reduces in summer and autumn season, whereas peak period of flower produce in winter and rainy season. After getting maximum flowers in a season, plant to be given rest for one month with restriction of watering and feeding. First plucking of flower will start after 18, 15 and 12 days of flower bud initiation during winter, rainy and summer season respectively. Gerbera plants are produce flower in polyhouse up to 4-5 years, but economically profitable harvest can be obtained flowers up to 36-40 months. The flower production is second year and third year was obtained almost twice than that of the first year. It has been estimated that, on an average 200-250 nos. of flowers can be obtained/sq.mt./year in polyhouse cultivation.

Field: Natural Ventilated Polyhouse

A well equipped poly house is used for a field of gerbera, where all instruments are set up under mechanized control system and regulated based on plant requirement. Such type of poly house should have drip and sprinkler system for irrigation, fertigation and PPC application. Besides these, others facilities are also attached in this house like fans, foggers, garnets, shade net, natural ventilation with insect proof net etc. for controlling light, temperature and humidity. For harnessing of efficacy and smoothly functioning of all systems, regular checking followed by repairing and maintenance of all instruments are very much essential. Hell (1992) reported the effect of different percentage (30, 50 and 70%) of shading on growth of gerbera, un-shaded plants showed significant growth depression but 30% shaded plants produced better quality cut flowers. Shurti *et al.* (2004) studied the performance of thirteen cultivars of *Gerbera jamesonii* ('Charmander', 'Dalma', 'Diablo', 'Francella', 'Goldengate', 'Goldflor', 'Magnum', 'Ornelia', 'Rosalin', 'Sangria', 'Savannah', 'Sunanda' and 'Vino') under shed net during 2001-02 and found that cv. 'Charmander' was tallest (45 cm) with longer floral life span (16.6 days), while cv. 'Savannah'had highest number of leaves per plant (31.20), leaf area (264.43 sq.cm), stalk length (63.20 cm), number of flowers per plant (12.11), but cv. 'Sangria'had highest flower diameter (12.50 cm) and cv. 'Vino' showed longest vase life (15.16 days). Cv. 'Dalma' and 'Goldflor' showed least earliest first flower bud initiation. Another experiment has conducted under polyhouse during 2002-03 on *Gerbera jamesonii* employing eleven cultivars ('Xenia', 'Kozak', 'Torro', 'Gold Disk', 'Masai', 'Yellow Venus', 'Lady', 'Harlekiju', 'Sphinx', Carrousel' and 'Pinta') by Dalal *et al.* (2005), they reported that the vegetative growth and flower yield were most promising in four cultivars ('Xenia', 'Kozak', 'Masai' and 'Yellow Venus').

Climate

This plant is very much sensitive to extreme temperature, light and humidity. During planting, plant requires day (22-25°C) and night (18-20°C) temperature for proper establishment on beds, but others stages of development from vegetative growth to flowering, plants are preferring temperature during day and night at 20-25°C and12-16°C respectively. It grows well in winter season in open condition, but special care should be taken during others season, especially in summer, when prevailing temperature arrived above 35°C and it is difficult to plant survive. For overcoming these situation foggers, fans and garnets are to be kept in functioning stage from 10am to 2.0 pm. Similarly when outside temperature goes down below 12°C, then side ventilation should be closed during whole day. Generally in normal condition, side ventilation kept open from 6.0 am to 6.00 pm and closed in rest of the time in a day. The relative humidity of 80-85% and

14 hours (35,000-40,000 lux) of day length is beneficial for plant growth. Light intensity during summer to be reduced by providing additional shade (30-35% green shade net) inside the polyhouse. During summer months, alternate day of hand force watering on the beds in the morning to be done, which improve freshness of the plant and also reduce insect population as well as protect the plants against dry rot in the polyhouse.

Growing Media Preparation and Planting

Gerbera plants are grown very well in organic base growing media. It is prepared with sandy loam soil along with organic matter (Coco-peat /FYM /Vermi-compost) in the ratio of 70:30 and mixed thoroughly, while growing media have adequate porous, well drained and retained sufficient moisture. For enrichment of growing media, neem cake 1kg/sq.mt is beneficial for plant growth. When loamy or clay soil are used in the growing media, then 10 and 15% of sand to be mixed in the growing media respectively for better aeration. After mixing of growing media, bed should be prepared with proper spacing like bed height (45cm), width of bed (upper surface: 60cm and lower base: 67cm) and space between the beds (30cm), but length of the bed depends on size of the green house. As soon as, when beds are prepared, then growing media should be disinfections with a treatment of Formaline (@ 100ml/lit. of water) through drenching on the beds (one sq.mt area) followed by mulching with plastic cover for 7 days and then thoroughly flush the beds by plan water (@ 100lit./sq.mt) and wait for two weeks for planting. Upper surface of the growing media up to a depth of 15cm again enrich by adding with some others nutrients (SSP @ 250gm/sq.mt + $MgSO_4$ @ 5gm/sq.mt. + Biozyme granules @ 20gm/sq.mt + Humid granules @ 20gm/sq.mt.) and after final bed preparation, do not walk on the beds. Tissue culture plants with root ball to be planted before or after rainy season, but during planting, plant to plant and row to row distance should be maintained with 30x30 and 37.5x37.5cm respectively followed by irrigation. But one thing should remember that crowns to be kept 1 cm above ground level during planting and when plants are established on the beds, plants are lightly put down, while crown to be set up finally, just above the ground level. After planting every day morning light irrigation to be provided and 80-90% humidity to be maintained up to 4-6 weeks to avoid desiccation of the plant. It has been estimated that 7 nos. of plants/sq.mt. and 6000 nos. of plants can be accommodated in 1000 sq. mt .area under polyhouse. As regards growing medium for gerbera. An experiment on gerbera cv. 'Bora Bora'was grown in pots in a perlite or 3:1 mixture of peat: perlite (V/V), plants grown in the mixture had a 20% greater leaf mineral content and 80% higher cut flower yield than those grown in perlite alone reported by Rea *et al.* (1999) whereas, Maloupa and Gerasopoulos (1999) grew four cultivars of gerbera

('Fame', 'Party', 'Regine' and 'Ximena') under plastic green house using perlite, zeolite, sand and rock wool as substrate and found that after 17 months period perlite medium give highest yield for all varieties and Ozcelik *et al.* (1999) reported that after 15 months of gerbera growing in polyhouse under different growing medias (perlite, peat, pumice and rock wool) either alone or in combinations yielded 59.31 number of flowers per plant from the plant grown in peat + pumice (1:1, V/V) followed by peat (57.71 nos of flowers per plant). Fiorenza and Paradiso (2000) noticed that orange colours of cultivars of gerbera produces highest number of flowers per plant (11.0) as compared to other colours cultivars, when plants grow in 70% peat + 30% perlite (by volume) as substrate. Whereas, under naturally ventilated polyhouse the plant of gerbera var. 'Sangaria' vegetative growth was best in coco-peat alone and flower quality with respect to head and disc diameter, number of ray florets and stalk length was superior in pots, when plants grow in the combination of peat and vermicompost (1:1, V/V) reported by Barreto and Jagtap (2002).

Nutrients Management and Watering

There is close relationship between fertigation and watering to the gerbera plants, because of fertilizers and micro-nutrients are applied with water through drip. AICRP-Floriculture scientists of BCKV at Moundari Farm observed during 2014-16 water soluble fertilizers like NPK: 19-19-19, NPK:13-0-45, NPK: 12-61-0 and micronutrients(CAN, Borax, $CuSO_4$, $MgSO_4$) are best for growth and flowering of gerbera. During vegetative growth up to three months after planting, plants were fertigated with NPK: 19-19-19 @ 1gm/lit. of water weekly. When plants are showed to start flower bud initiation, then fertigation schedule slightly change in first and third weeks (NPK: 19-19-19 @ 1gm/lit. of water + NPK:12-61-0 @ ½ gm/lit. of water + CAN @ ½ gm/lit. of water + $MgSO_4$ @ 0.10gm/lit. of water) as well as second and fourth weeks (NPK: 19-19-19 @ 1gm/lit.of water + NPK:13-0-45 @ ½ gm/lit. of water + Borax @ 1gm/lit. of water + $CuSO_4$ 1gm/lit. of water). All fertilizers and micronutrients were applied separately at alternate day. Before and after use of fertilizers through drip, growing media should irrigate by plan water through drip (5-6 minutes). An experiment was conducted during 2000-01 under poly tunnel on gerbera cv. 'Ornella' 15gN+20gP+20gK/sq.m resulted maximum vegetative growth as well as produced good quality and quantity of flower reported by Gurav *et al.* (2002) and the next year (2001-02) gerbera var.'Popular' treated with NPK fertilizer at 10, 20 and 30g/ sq.m of each, NPK at 30:10:20g/sq.m had improved on the plant height (45.69 cm), leaf area per plant (3435.67 sq.cm), least days to first flower appear (84.6 days), number of flowers per plant (67.01), flower size (912.15 cm), stalk length (42.79 cm), vase life of flower (26.35 days) and sucker per plant (30.40) by Hunmilli and Paswan (2003). Sujatha(2002)observed that the application of 80% recommended level of water soluble or straight fertilizers through fertigation was most effective for improved growth and flower production in gerbera. Nayek et al (2005) observed that a mixture of NPK at 4:2:2 g/pot was found to be the best performer in respect of number of plantlets(3.7)/pot, days to flower bud formation(137.20days), number of flowers/plant(15.60) and diameter of flower(8.92cm). Barad *et al* (2010) observed that for maximum growth, flower yield and quality of gerbera flowers Cv. Sangaria under net house conditions, when plants were fertilized with 20:10:20gm/sq.mt N:P:K. The optimum leaf nitrogen ranged 2.54to 2.87%, phosphorus from0.13 to 0.61 and potassium from 3.37 to 4.37% (Anjaneylu, 2008) and whereas Chowdhuri (2008) reported that NPK content in the leaf of gerbera (1.45, 0.18, 2.48%)is best for flower production. It has been found that after 3-4 months of continuous flower harvesting, plants should be brought under rest for one month or plants are to be kept under rest during summer or there is no any local festival. During this period, racking of beds on upper surface to be done and also all dry as well as infected older leaves are to be removed. Sometimes beds are requiring little bit of repair followed by application of vermicompost @ 1kg/sq.mt. and this process till continued up to three years or last harvesting. Watering is influences on the growth and yield of plants and it has been found that excess water and low level of water used is harmful to the plants. In the same way, quality of water in terms of P^H(6.5-7) and EC (<0.7ms/cm), TDS (<450ppm) and hardiness (<200ppm) are to be maintained for obtaining better yield. Up to one month from planting, over head irrigation through sprinkler is to be provided and after that drip irrigation to be followed. One dropper should be used for one plant. Moring time irrigation is the best in a day and tries to keep soil moist moderately as thumb rule. Always fresh water use is beneficial to the plant in terms of freshness of plants as well as reduces diseases. Growing media should be tested (P^H, EC, TDS) every alternate three months of interval.

Application of Growth Regulators

Growth regulators, especially growth promoting substances played a vital role in growth and development of gerbera under polyhouse cultivation. Farina *et al.* (1989) applied GA_3 at 100ppm as foliar spray on gerbera at monthly interval and observed that cut flower diameter was increased. Another experiment was conducted by Lee and Lee (1990) in potted gerbera single drench with growth retardant like paclobutrazol at 0.25-0.50 mg/pot resulted the reduction of peduncle length, leaf area, foliage height and width of leaf. Krishna Pal Singh (2001) reported that spraying of GA_3 or NAA (200 ppm) at 5-6 leaf stage significantly increased plant height and number of flowers per plant, but spraying of B-9 at

2500 ppm at 2-6 weeks after transplanting resulted not only in reducing scape length but also reduced surface. Foliar application of from January to May was best for growth, maximum nos. of cut blooms with stalk lenth as well as flower size meeting the global standard.

Plant Protection Measures

Very careful supervision is very much an essential for gerbera cultivation under natural ventilated house, anytime can happen anything creates a big problem. Sometime, problems are not found from outside, which is internally affected to the plants, resulted to produce very inferior quality of flower.

When plants are showing yellowing in older leaves causes nitrogen deficiency, but such types symptoms found in new leaves also, causes by manganese (Veins remain green) and calcium(extreme yellowing) deficiency. Browning discoloration is also one of the problems in gerbera plant due to phosphorus deficiency. Marginal necrosis, interveinal chlorosis on older leaves, interveinal chlorosis on younger leaves and complete chlorosis on younger leaves are caused by deficiency of potassium, magnesium, iron and cupper respectively. In very few cases, younger leaves become blackish; this is related to boron deficiency. All major and micronutrients are playing a vital role in quality of gerbera flower production and their application already been discussed earlier.

Among the insects and pest, Red Mites (Suck the sap from the lower surface of the leaves, causing development of bown spots on the lower surface and also found marginal drying) and Cyclamine Mites (Leaves are affected like older leaves are curled-up, younger leaves become uneven formed and leathery, and flowers are deformed, petals are missing, inward curling and discoloration) are very much harmful to this plants almost round the year. Anytime they can damaged the plants, so routine spraying to be done at fortnight interval with one of the chemicals (Dicofol @ 2.5ml /Spiromecifen @ 1ml / Diaphenthiuron @ 1gm) at an alternately. Besides these two insects, others insects like Leaf Minor (control: Cartap hydrochloride @ 1gm), Thrips (control: Fipronil @ 1,5ml), White Fly (Imidacloprid @ 0.5ml) and Catterpiller (Control: Cholopyriphos @ 1gm/plant) are also damaged this plant. The doses mentioned above are to be applied per liter of water.

There are so many diseases affected to the plants are Crown rot (wilting of gerbera, control: Copper oxychloride @ 1.5gm), Root rot (Initially dropping of younger leaves and finally plant wilting, control: Carbendazim @ 2 gm), Fusarium rot (Blackening of crown portion and plant appear brown discoloration, control: Topsin-M @ 1.5gm), Leaf spot (Black circular spot appears on leaves, control: Carbendazim @ 2 gm), Powdery mildew (White powder developed on leaves, control: Dinocap @ 0.4ml), Botrytis(Gray spot develop on flower petals, control: Dithane M-45 @ 0.5gm), Bacterial blight(Yellowing oily spot on leaves, control: Streptomycine @ 50mg). As early as possible all chemicals are to be sprayed, as when required, but during spraying do not mixed along with nutrients or any fungicides. The doses mentioned above are to be applied per liter of water.

Harvesting and Yield

Harvesting of flowers depends on location of markets. Generality flowers are harvested, when 2-3 whorls of stamens have entirely been developed or outer two rows of disc florets are perpendicular to the stalk. Early morning or late evening is the best time for flower plucking. During harvesting, one thing should remember that, flower stick plucking is better than cutting. As soon as harvest the flowers, cut end of the flower should dipped into the water and kept 15-16°C temperature with high humidity for four to five hours before transportation. After two months of planting, when plants having 15-16 nos. of leaves, flower buds initiation will take place. So, first plucking of flower stick will start after 18 days of flower bud initiation and harvesting of flower will continues up to 36-40 months. The flower production is second year and third year was obtained almost twice that of the first year. It has been estimated that, on an average 200-250 nos. of flowers can be obtained/sq.mt./year.

Post Harvest Management

Presently it is very much needed for proper marketing, because of several unavoidable circumstances happened during transportation like labour strike, road strike, flight cancelled etc. It has been recorded through research that, 30-35% of total production will lost due to lacking of post harvest treatment facilities in the production unit. So, for improving the post harvest life of flower, it should be immediately immersing for 24 hours in a solution containing with HQS(200mg/lit. of water) + $AgNO_3$(50mg/lit. of water) + 5% sucrose.

References

Anjaneyulu, K. (2008). Diagnostic leaf nutrient norms and identification of yield limiting nutrients in gerbera grown under protected conditions using DRIS. Indian Journal of Horticulure, 65(2):176-179.

Barad, A.V., Nandre, B.M and Sonwalkar, N.H. (2010). Effects of NPK levels on gerbera Cv. Sangaris under net house conditions. Indian Journal of Horticulure, 67(3):421-424.

Barreto, M.S. and Jagtap, K. B. (2002). Studies of polyhouse gerbera substrate. In floriculture research trend in India. Indian Society of Ornamental Horticulture (2002), pp.173-176.

Chowdhuri, T.K. (2008). Standardization of agro-techniques for flower production in gerbera, anthurium, gladiolus and spathiphyllum in polyhouse, Ph. D Thesis, Univ. of Calcutta., Calcutta, pp.158.

Dalal, S. R., Gonge, V.S., Mohariya, A. D. and Anjue, A. (2005). Performance of different varieties of gerbera under polyhouse conditions. Advances in Plant Science, 18(1), pp. 269-272.

Das, P. and Singh Samanta, P.K. 1999. Gerbera. Floriculture and Landscaping, edited: T. K. Bose; R.G. Maiti, R.S. Dhua and P. Das, Naya Prokash, pp.513-519.

Farina, E., Paterniani, T. and Volpi, L. (1989). Effect of GA_3 treatments on flowering of gerbera grown for winter production. Acta Horticulture, No. 246, pp.159-166.

Fiorenza, S. and Paradiso, R. (2000). Evaluation of gerbera cultivars in an open soil less system. Colture Protette, 29(9):125-128.

Gurav, S. B., Singh, B. R., Katwate, S.M. and Yadav, E.D. (2002). Standardization of fertilizer dose for flower production of gerbera under protected conditions. Indian Society of Ornamental Horticulture (2002):313-314.

Hell, B. Ter; Ludolph, D. and Kleiber, B. (1992). Controlled culture of gerbera in summer and winter. Gartenbau Magazine, 1(6):64-66.

Hunmili Terangpi and Paswan, L. (2003). Effect of NPK on growth and flowering of gerbera. Journal of Indian Horticulture, 6(1):71-72.

Krishna Pal Singh. (2001). Agro-Techniques for gerbera cultivation in India. Floriculture Today, February-2001, pp. 9-13.

Lee, O. P. and Lee, J. S. (1990). Effect of ancymidol and paclobutrazol on growth and flowering of potted gerbera. Journal of Korean Society for Horticultural Science, 31(3):300-304.

Maloupa, E. and Gerasopoulos, D. (1999). Quality production of four cut gerberas in a hydroponics system of four substrates. Acta Horticulture, 491:433-438.

Nair Sujatha A., Singh Vijal, Sharma, T.V.R.S.(2002). Effect of plant growth regulators on yield and quality of gerbera under Bay Island conditions, Indian Journal of Horticulture, 59, Issue-1:100-105.

Nayek, D., Mandal, T. and Roychowdhury, N.(2005). Effect of NPK nutrition on growth and flowering of Gerbera jamesonii L Cv.Constance. Orissa J.Hort., 33(2):11-15.

Ozcelik, A; Besiroglu, A. and Ozgumus, A. (1999). The use of different growing medias in green house gerbera cut flower production. Acta Horticulture, No. 491:425-432.

Rea, E., Pierandrel, F., Cantone, P. and Maletta, M. (1999). Mineral composition of soil less culture gerberas. Culture Protette, 28(6):71-75.

Shruti Wan Hede; Golliwar, V.J., Shital Dagwar; Manjusha Atthawle and Nisha Bhaladhare. (2004). Performance of different varieties of gerbera under shade net. Journal of Soil and Crops, 14(2):383-387.

Sujatha, K. (2002). Fertigation studies in gerbera under low cost greenhouse, Ph D Thesis, Univ. Agric Sci., Bangalore, pp37-50.

Photographs showing different varieties of gerbera (*Gerbera jamesonii*) growing under Natural ventilated polyhouse

6

Protected Cultivation Technology of *Alstroemeria* as Commercial Flower Crop

M.K. Singh

Division of Floriculture and Landscaping, Indian Agricultural Research Institute Pusa, New Delhi-110012

Introduction

Alstroemeria also known as the lily of the Incas, Peruvian lily or Inca lily, has been grown mainly as a cut flower crop. Recently, it has become popular as a garden flower and also as a potted flowering plant. The plants produce beautiful large inflorescence in many different colours *viz*. lavender, orange, yellow to dark yellow, white, pink, red, purple and bi-colors. Flowers are characterized by black dots at the base of the petals and throats. The cut flowers have a long post harvest life, up-to two weeks. These valuable characteristics have made *Alstroemeria o*ne of the top cut flowers at the Dutch Flower Auction Centers. The plants produce high yield and possess an ever blooming habit after flower initiation has occurred but in general most cultivars flower best during spring and early summer. The straight, un-branched shoots emerge from underground rhizomes which also produce enlarged storage roots. It is gaining popularity in Indian flower market due to long stem flowers and prolonged vase life. Recently flower growers have shown great interest about its production as a cut flower crop.

Botanical Name	:	*Alstroemeria* L.hybrids (Huxley *et. al.*, 1992)
Genus	:	*Alstroemeria*
Family	:	*Alstroemeria*ceae
SubClass	:	Monocotyledonae

History and Botany

The center of origin of *Alstroemeria* is found in Latin American, especially countries like Chile, Peru and Brazil. The name *Alstroemeria* is given by Linneaus after the Swedish naturalist klas von Alstroemer (1736-94). It has been found from dry, warm regions to moist, cool, high elevation habitats and from 26^0 to 40^0C south latitude (Bridgen, 1997). It is an herbaceous perennial consist of multistemmed rhizome from which aerial shoots and fibrous roots develops. Later these fibrous roots become thickened storage roots which are called "Radices medullosae". These roots are white, fleshy, very brittle and dense. The flowers of *Alstroemeria* are zygomorphic, in a terminal, simple or compound, rarely solitary, tepals 6, free, clawed, the 3 inner tepals narrower and longer than the outer; stamens 6, declinate, attached to base of tepals, anthers basifixed, style slender, stigma 3-fid, ovary inferior, 3-celled. Inflorescence is cyme and each cyme is sympodially branched with upto four florets per cyme that open one after another. Stamens open and shed their pollens before the stigma becomes receptive (Traub, 1943). Leaves have parallel veins which are grey-green to dark green and mostly hairless on both sides. Leaf petioles twist 180^0 on the stems so that original underside becomes the top side (Bayer, 1989). Stomata's are on the upper surface as well as on the underleaf surface. (Healy and Wilkins, 1985; Heins and Wilkins, 1979). Fruits are many seeded capsule.

Cytology and Breeding

The chromosomes of *Alstroemeria* are very large. The basic chromosome number of *Alstroemeria* is n=8 .The somatic chromosome numbers of wild *Alstroemeria* species have been determined by Karyotype analysis and it is 2n=2X=16. (Srasburger 1882; Taylor 1926; Hang and Tsuchiya 1988; Stephens *et al.*, 1993).

Breeding

Primary objectives in breeding programs for *Alstroemeria* are to develop a hybrid plants that will possess the traits such as attractive flower colours, flower longevity, fragrance, compact growth habit and virus and disease resistance. A lot of cultivars have been developed by interspecific and intraspecific breeding. Presently, biotechnological approaches are used to improve *Alstroemeria* strains. Interspecific hybrid plants have been produced by ovule cultures. By improving certain culture techniques, sexual incompatability was overcome in some cross combinations. Plant regeneration through callus culture has been reported. Particle bombardment and *Agrobacterium*-mediated procedures were applied for genetic transformation and some transformed plants with marker genes were produced. In *Alstroemeria*, isolation of egg cells and zygotes from ovules has been attempted in order to develop an *in vitro* fertilization technique (Hoshino *et al.* 2006).

Species and Varieties

The genus *Alstroemeria* comprises more than 60 species. Hofreiter *et al.* (2006) reported that there are 39 species present in Brazil, 33 species in Chile, 10 species in Argentina, 2 species in Peru and 1 species in Bolivia. According to Van Scheepen (1991) the classification and descriptions of some of the most common species of *Alstroemeria* are as follows:

A. Aurea Graham (*A. aurantiaca* D.Don)

Species from southern Chile. Stems upto 1 m, leaves 7-10 cm long, lanceolate, subsessiles, often twisted to expose their grey-green undersides, fleshy roots, suited for warm porous soils, very sensitive ,needs protection with a covering of leaves in winter. Umbels 3-7 rayed with 1-3 flowers per ray, tepals 4-5 cm, flowers bright orange or yellow with purple stripes, outer tepals broadly ovate,obtuse, tipped green, inner tepals acute, upper part spotted and flecked red, lowermost tepal sometimes unspotted.

Cultivars : Orange King-Orange, Moerheim-Orange, Dover Orange-Deep Orange, Lutea- Yellow

Other cultivars: Angustifolia, Flava, Major, Rubra, Splendors

A. brasiliensis Spreng.

It is native of Brazil. Stems up to 120 cm, leaves 5-10 cm long, linear, erecto-patent, glabrous, those of sterile stems 7.5-10 cm, lanceolate, Silver grey midrib. Umbels usually 5-rayed with 1-3 flowers per ray, tepals 4 cm, red-yellow, inner tepals flecked brown and stamens shorter than tepals.

A. Chilensis hort.

Chilean species from the Termas de Chillen, Plants up to 75 cm. Leaves scattered, narrowly obovate and margins minutely ciliate. Umbels 5-6 rayed, flowers 2 per ray, tepals pale pink to blood red, upper inner tepals yellow striped red-purple.

A. haemantha Ruize & Pav

This species originates in Chile. Stems 60-90 cm, leaves 7-15 cm, narrowly lanceolate, subsessiles, grey-green underside, margin hairy. Umbel 3-15 rayed; flowers 1-4 per ray; tepals 4-5 cm, oblanceolate-spathulate, narrow, orange to deep vermilion, outer tepals tipped green, upper inner tepals yellow-orange striped purple, lowermost tepal orange to orange-red, striped dark red. Cv-Rosea.

A. hookeri Lodd.

Resembles *A.pelegrina* but tepals pink, upper inner tepals blotched yellow and flecked red-purple and native Peru.

***A. ligtu* L.**

Alsroemeria ligtu is also called St Martin's flower. Stems 45-60 cm long and leaves 5-8 cm, narrowly lanceolate to linear- lanceolate, erecto-patent. Umbels 3-8 rayed; Flowers 2-3 per ray; pedicles 5-7.5 cm; tepals white to pale lilac to pink-red, obovate, upper inner tepals usually yellow, spotted and streaked white or purple or yellow and red; stamens shorter than tepals and native Chile and Argentina.

A. pelegrina L.

This species originated in Chile and is used extensively in breeding. Stems 30-60 cm and leaves 5-8 cm, lanceolate, glabrous. Flowers solitary or in 2-3 rayed umbels; tepals 5 cm, off-white flushed mauve or pink with with a darker central zone, inner teplas yellow at base, flecked brown or marron. Cultivars Alba-Flowers white, faded green, inner upper tepals spotted green and often spotted yellow and cv Rosea with deep rose flowers.

A. psittacina Lehm.

A species from Brazil that may be similar to *A. pulchella* L.f. Stems up to 90 cm, spotted mauve. Leaves up to 7.5 cm, lanceolate, glabrous. Umbels 4-6 rayed; tepals 4-4.5 cm, green overlaid with dark wine red, spotted and streaked red or maroon; stamens as long as tepals.

A. pygmaea Herb

A species from Argentina that may be similar to *A. aurea and* plants up to 20 cm only. Stem subterranean, producing a group of leaves up to 2.5 cm, grey-green; flowers 5 cm long, yellow, inner tepals and spotted red.

A. versicolor Ruiz & Pav

It is native of Chile. Resembles *A. psittacina* but smaller. Stems 20-60 cm, leaves linear, around 2.5 cm. Umbels 2-4 rayed; tepals 2.5 cm, pale yellow or orange with purple spots. Inner segments spotted and streaked.

A. violacea Philippi.

Chilean species with stems 50-100 cm. Leaves 6-9 cm, ovate-oblong, spreading, scattered, glabrous. Umbels 3-6 rayed, flowers 3-5 per ray; tepals 3.5 -5.5cm, bright violet, outer tepals obovate, blunt, shortly cuspidate, inner tepals oblong-acute, white at base, spotted purple, sometimes flushed orange; stamens shorter than tepals.

A. angustifolia Herbert

Native to Chile, stems 30-60 cm, leaves narrowly lanceolate, flowers pale pinkish, 7-8 cm across, the upper lateral segements finely spotted, with a yellow band.

A. pulchra anon.

It stems up to 45 cm, flowers white to light grayish-pink or soft lilac that are spotted yellow, red or purple.

A. zoellneri E. Bayer

Native to Chile in Aconagua and Santiago provinces, stems 30-60 cm, leaves linear, around 2.5 cm, flowers 5-6 cm across, the outer segment ovate, acuminate with a dark spot near the apex; the inner laterals recurved with a yellow band with purple streaks.

A caryophyllaca

Fragrant species originated from Brazil, stems up to 45 cm, flowers with rose sepals and lower petals, upper petals white in centers.

The choice of *Alstroemeria* cultivars are determined by many factors. A good variety should meet the following requirements: large flower buds that show complete colour, good bud formation, large flowers, strong colours, strong in handling and good keeping quality. A limited flower length and a high flower yield are positive characteristics which help to increase profitability. On the basis of a study at CSIR-IHBT, Palampur (H.P.) few cultivars *viz.* Tiara, Pluto, Cindrella, Capri, Rosita, Serena, Aladdin and Amor are recommended for commercial cultivation in hilly regions (Singh, 2006a).

Recommended Alstroemeria Cultivars

A few selected types of cultivars are given below:

Butterfly Type: Jessica (deep pink), Claudia (pink), Bombay (bright pink), Symphony (red) etc.

Hybrid Type: Bounty (dark pink), Amber (orange), Disco (lilac), Pluto (orange), Sylvia (creamy white) etc.

Aurantiaca Type: Aladdin (lemon yellow), Gold finger (yellow), Mandarino (orange) etc.

Brasiliensis: Bambi (orange), Petit Rouge (red), Visa (red) etc.

Propagation

Alstroemeria is vegetative propagated by rhizome division or through micro-propagation. Asexual propagation allows plants to grow true to type. Divide bed growing plants of *Alstroemeria* at least every second and third year, depending on the variety and growth characteristics. About one to two weeks prior to division, plants should be severely pruned, leaving only the youngest 15 to 20 cm long shoots. Roots of the rhizomes grow 30-40 cm deep thus dig the roots properly to get the feeding roots along with growing points for making division. The rhizomes with roots and shoots should be planted separately. The crop is also propagated by seed for cut flower production and to develop new hybrid varieties.

Rhizomes of *Alstroemeria* cv 'Pluto'was planted under 35% shade nets to study the effect of number of storage root and nitrogen fertigation on growth and rhizome production parameters under Kashmir conditions. It has been found that maximum values in terms of percent established plants (61.39%), number of vegetative shoots (4.78), weight of the rhizome cluster (13.11 g), number of rhizomes developed (2.38), length of the longest rhizomes (5.50 cm), number of new storage roots as well as fibrous roots (5.57 and 6.31) and the propagation coefficient (31.18) were recorded with the highest level of storage roots 2-4. Among the nitrogen levels tested, 200ppm recorded highest values in terms of percent established plants (63.34%), number of vegetative shoots (4.45), weight of the rhizome cluster (11.61 g), number of new storage roots as well as fibrous roots (4.82 and 6.14) and the propagation coefficient (24.32) over lower levels of nitrogen tested in the study. (Singh *et al*, 2010a). In an experiment to study the effect of no. of storage roots and benzyladenine on growth and rhizome production, it has been found that maximum values in terms of percent sprouting, percent established plants, number of vegetative shoots, weight of the rhizome cluster/plant, number of rhizomes developed, length of the longest rhizome, number of new storage roots, number of new fibrous roots and propagation coefficient were recorded with highest level of storage roots 2-4 and benzyladenine, 60ppm BA recorded highest values in terms of percent sprouting, number of rhizomes developed, length of the longest rhizome, number of new storage roots, fibrous roots per plant and propagation coefficient over other levels of BA(0, 20 and 60 ppm) (Singh *et al.*, 2010b).

Climate

Alstroemeria prefers location having cool temperature, free from water logging and strong winds. It does not prefer direct sunlight and can be grown successfully in cool place under poly house conditions. Cooling system is required in the place having higher temperature. The crop is most comfortably suited to a relative humidity between 65-85%. By CO_2 dosing 900 ppm, the flower quality is better, the time of flowering is early and the production of flower is increased.

Light

Sufficient light (5000 ft. candles) is important to prevent bud abortion which results in strong and heavy quality flowering stems. Long day speeds up the development of the flower, but decreases the production of shoot. Short days decreases flower production, suppress rhizome formation and growth. The optimum photoperiod appears to be 12-16 hours, although day length more than 16 hours induces earlier flowering but decreases total flowering shoot production, so it is not recommended for commercial production.

Temperature

Temperature is very important for *Alstroemeria* cultivation and flower induction occurs at a soil temperature of 14-20^0 C. Temperature higher than 24^0 C should be avoided, as high temperature inhibits the flower induction. However, it has been observed that *Alstroemeria* shows differential response to temperature. Optimum air temperature for its cultivation should be 12-16^0 C during night and 18-20^0 C during the day. Prolonged air temperature over 24^0 C may decrease or stop flowering. *Alstroemeria* can tolerate temperature as low as 5^0 C in winter.

Soil

For the best results the most important factors are the structure and the water retention capacity of the soil. In general, soil with a pH of 5.5 to7.0 and electrical conductivity (EC) less than 1.0 is best suited for growing the crop. *Alstroemeria* shows a much quicker and luxuriant growth in peat and sandy loam soil than in clay soil. In case of clay soil, it is advised to mix sand and well decomposed farm yard manure (FYM) to make it porous. Good drainage is vital to avoid root-rot and fungal diseases.

Land Preparation

For cut flower production, ground beds should be made well pulverized. Soil should be dug to a depth of 30-40 cm deep to allow the roots to grow during the two year production cycle. Beds should be 1.0 to 1.2 meter wide and paths should be minimum 50-60 cm between the beds. Before planting, mix the well decomposed dry organic manure @ 5.0 - 8.0 kg per sqm in cultivated area.

Cultivation Practices

Planting and Plant Density

Alstroemeria can be planted year round, but planting time can differ because of the required flowering time and seasonal problems like high temperature. It is planted in two rows parallel to each other. Depending upon variety the distance between each plant varies from 30-50 cm. It is advisable to keep the planting density 4-6 plants per sqm net area. Plant rhizomes at 20 cm deep and spread out roots properly in the pit. *Alstroemeria* can also be grown in larger pots for decoration. One plant per 13 cm pot or 3 plants per 25 cm pot are planted.

Depending on light intensity and temperature, it takes 3-5 months for the first flower to bloom after planting. When it is planted during October – November in polyhouse conditions of Palampur, Himachal Pradesh, flowers rapidly from March to June and produce the greatest number of generative shoots.

Control of Growth and Flowering Including Shading, Thinning and Support

Shading

Shading is necessary to protect the crop and soil from the sun. Green shading net of 25% are recommended on the polyhouse roof. It is important to find a good balance between shading for cooling and to keep optimum light intensity in the polyhouse.

Support

Alstroemeria plants grow 50 to 150 cm tall depending upon the cultivars. So, staking is required to hold the plants. It also keeps the flowering stem erect. It is recommended to use 2 to 3 layers of wide nets. It should be spread in the beds as soon as the rhizomes start sprouting. The height of supporting net should be raised as the plants grow in height. The lower most net should be fixed at 30 cm above the ground level.

Thinning

Thinning of undesirable shoots promotes the production of new shoots and improves the quality of flowers. It is advisable to remove weak and blind (non-flowering) shoots on a monthly basis in late autumn and winter from the production plot. Investigation was conducted on 1-year old plants of *Alstroemeria* cv 'Serena' on production of cut flowers under polyhouse conditions by thinning 0, 30, 60 and 90% of vegetative shoots on monthly basis during flower production period (March-July). The maximum flowering stems per plant/year and length of flowering stem were observed 75 and 91.37 cm respectively by 30% thinning of vegetative shoots followed by 60% thinning (61.50 and 87.5 cm respectively). Thinning of 30% vegetative shoots gave better response for 'A' grade of flowering stems 38 nos./plant/year followed by 60, 90% thinning and without thinning of vegetative shoots (34, 32, and 24 nos./plant/year respectively) (Singh *et al.*, 2006b).This recommendation support the view of Healy and Wilkins (1991) that stem removal by pulling technique stimulates bud break and formation of new aerial shoot. Without thinning of vegetative shoots, plants produces less number of flowering stems as compared to other treatments.

Water Quality and Watering

Alstroemeria is sensitive to salt. It is, therefore, of great importance that good quality water is used for the better growth of plants. Rain water is highly suitable for its cultivation. The salt content in the irrigation water should be less than10 micro mol per liter. Top 30 cm of soil should be continuously kept moist because the development of root takes place up to this depth. *Alstroemeria* crop can evaporate 3 to 6 l of water per sqm area per day, if sufficient water is not given to the crop, the flower production declines. Newly planted crop should not be over irrigated.

Fertilization

Before planting soil analysis should be done to ascertain the fertility level of the soil. It requires fertile soil with high nutrient levels once the plants are established. Regular fertilization with 300 ppm N_2 and 300 ppm K_2O through Ca $(NO_3)_2$ and KNO_3 is recommended per plant per week for good growth and quality flower production. Nitrogen should be provided in the form of nitrates under cool growing conditions.

Use of Growth Regulators for Height Control

Various concentrations of Alar (Daminozide), Chlormequat (Cycocel) and Paclobutrazol (Cultar) as spray, drench and combination of both drench and spray applications were tried on two pot grown cultivars of *Alstroemeria* i.e. Selection No.14 and Riana to study their effect on plant height, spread and other flowering parameters under glasshouse conditions at Mashobra (H.P.) were evaluated during two successive flushes. Both the cultivars responded favourably to retardant treatments. Maximum 31.5 percent height reduction in the first flush was recorded with Alar-1500 ppm spray whereas a combination drench and spray application of Alar-1500 ppm resulted in maximum 28 percent height reduction in the second flush. Cultivar 'Riana' in the second flush resulted in maximum height reduction and also produced maximum spread. Cultivar' Selection

No.14' took minimum days to bud formation, produced greater number of cymes per inflorescence in second flush and responded with a better pot presentability than cv 'Riana' which remained in flowering for maximum duration in both the flushes. Significant reduction in days to bud formation was recorded with PP_{333} 50 ppm drench and spray in the first flush. A marked positive effect of PP_{333} 50 ppm drench and spray along with Alar 1500 ppm drench and spray was observed on number of cymes per inflorescence in the first flush whereas in the second flush, cycocel 1500ppm drench and spray resulted in maximum number of cymes per inflorescence. All the treatments significantly increased duration of flowering and pot presentability in both flushes, with spray application of Alar-1500 ppm coupled with drench and spray application of cycocel 1500 ppm and PP_{333} 50 ppm producing the best results. (Wazir *et al,* 2010).

Post-Harvest Management of Rhizome

Storage of Rhizome

An experiment was conducted to study the effect of low temperature storage duration (0, 4, 6 and 8 weeks0 at 2-4^{0}C in a cold store and thidiazuron application (0, 15, 30 and 45 ppm) on rhizome multiplication in *Alstroemeria* 'Alladin'. The rhizomes were soaked for 12 hours in thidiazuron (TDZ) solution and then given low temperature treatment of different durations. Planting was done in polythene bags filled with sand+soil+FYM in first week of September (autumn) and again in first week of March (spring). Maximum percent sprouting and percent establishment of plants was achieved when rhizomes were kept for 6 weeks in cold store. A concentration of 15ppm thidiazuron was best and higher concentrations had a negative effect on percent sprouting and percent establishment of plants. The number of shoots produced after intervals of 40, 60 and 80 days, were significantly increased by storage durations. Storage durations of 6 weeks were best for production of shoots; however, in spring no cold storage of rhizomes also showed similar response. The maximum numbers of shoots were produced when rhizomes were not treated with thidiazuron. The number of shoots decreased progressively with increase in thidiazuron concentrations. The shoot production was higher in spring as compared to autumn. Storage duration or thidiazuron treatments had no significant effect on stem diameter and chlorophyll content of the leaves. (Gupta *et al.*, 2009)

Flower forcing

Alstroemeria can flower round the year depending on the care and attention of the crop and also on the cultivars. When *Alstroemeria* planted in greenhouse in the month of October-November, it flowers from March to August in the next year. However, in the second year same plant produces flowering stems from

March to July. It has been observed that 30 to 35% flowers were produced during the month of May only (Singh, 2005).

Flower Harvesting

Flowering shoot emergence after 5 to 6 months planting, depending upon the cultivars and it takes further 25-35 days to flower. Flowering shoots should be pulled when the first floret are fully coloured and the majority are showing colour in the inflorescence. For long distance market, shoots may be harvested when the first floret is swollen and about to open. Flowers should be kept immediately in the fresh water after cutting the lower portion of shoots. It should be store at 2-4^{0}C temperature. Flowers and leaves are sensitive to ethylene and turns yellow after harvest. Sometimes cut flowers of *Alstroemeria* exuds a sap that can cause a skin irritation, so it is advisable to wear rubber gloves to prevent irritation.

In an experiment to study the effect of GA_3 and benzylaminopurine on the vase-life and post-harvest life quality of *Alstroemeria* , freshly cut flowering stems of cv. 'Alladin' were placed in a solutions containing 0, 25, 50 and 75ppm of GA_3, BA or GA_3 plus BA/l . Flowers held in these chemicals significantly delayed the onset of flower senescence as measured by days to 50% leaf yellowing(14 days) was found with GA_3(75 ppm) +BA(75ppm). These results indicate that GA_3 plus BA at 75 ppm has the potential to use as a commercial cut flower preservative solution for delaying flower senescence, prolonging the vase life and enhancing post harvest quality of *Alstroemeria* cut flowers (Dhiman, 2009).

Alstroemeria cut flowers held in pulsing treatment with sucrose (10%) + 8-HQC (300 ppm) + BA (25ppm) solution for 8 hours resulted in maximum vase life with best appearance. When it is stored for 24 hours at 4^0 C in sucrose (4%) + 8-HQC (200 ppm) + BA (10 ppm) solution and then held in fresh solution of same composition resulted in maximum vase life with best appearance of flower (Manikrao, 2007).

Yield

Yield of flowering shoots mainly depends upon the cultivars, plant spacing, growing conditions and cultural practices. *Alstroemeria* plant can produce 50-75 flowering shoots per plants per year under polyhouse conditions.

A study was conducted to evaluate the performance of nine *Alstroemeria* cultivars viz. Pluto, Alladin, Serena, Variety No.14, Capri, Tiara, Cindrella, Rosita and Amor under polyhouse conditions. The results indicate that maximum length of flowering shoot, diameter of flowering shoot, number of 'A' grade flowering shoots, number of cymes per inflorescence, number of florets per shoot and per cyme were recorded in variety No. 14. Flowers of variety No. 14 and Cindrella took 29.00 and 27.88 days, respectively for senescence on the plants and were superior to other cultivars. Variety No. 14 recorded the maximum vase-life (12.33 days) and was at par with Alladin (12.22 days) when placed in fresh water at room temperature. Minimum diseases incidence was observed in cvs Pluto, Capri, Serena, Tiara, Cindrella, Rosita and Amor. It was also observed that the maximum percentage of flower production was May (39.72%) and April (18.48%) in each cultivars. Maximum flower produced was in April in cvs Pluto, Alladin, Tiara and Rosita in March cvs Serena, Variety No.14, Capri, Cindrella and Amor. The maximum number of flowering shoots was produced by Serena and the minimum time taken for flowering by Alladin. The maximum length of primary floret was recorded in Rosita. The width of primary floret at full bloom stage was maximum in Capri. The maximum number of florets opened at a time was recorded in Cindrella. Minimum diseases incidence were observed in cvs Pluto, Capri, Serena, Tiara, Cindrella, Rosita and Amor. For round the year flower production, cultivars, Tiara and Rosita were found promising. This could be due to inherent genetic differences among cultivars. Based on the results, it has been suggested that introduction of all cultivars except Variety No. 14 (due to highly disease susceptible) can be done for the cut flower production in hilly regions. The grade of flowering stems classified according to the number of cymes/stem: grade 'A' with 5 or more cymes, 'B' grade with 4 cymes and grade 'C' 3 cymes) (Singh, 2006a).

Diseases and Pest

Diseases (Bacterial, Fungal, Viral and Physiological Disorder)

Alstroemerias are susceptible to number of fungal pathogens such as Botrytis, *Pythium* and *Rhizoctinia* spp., which attack the plants and flowers. One of the methods to prevent these diseases is soil disinfection. This is done after incorporation of organic materials in the ground and before planting. Disinfection of soil by steam controls ground fungus and nematodes. At the same time it also keeps the soil free of weeds for a longer period. It is also essential to provide adequate ventilation and spacing for plants.

Botrytis

In a *Botrytis* infected plant, brown colour spot are developed and visible on the flower petals during the period of high humidity. It can be controlled by providing adequate ventilation and keeping the crop dry during rainy season and spray of Mancozeb @ 2.0 g per liter of water at an interval of 10 to 12 days.

Root-rot

Root-rot is caused by the fungus *Pythium* which spreads in moist conditions. Infected plants lose the root parts and causes heavy loss to the flower production. It generally attacks the plants grown in heavy and compact soil. Root-rot can be controlled by sterilization of soil before planting, better air circulation and decreasing the moisture content of growing media.

Foot-rot

Rhizoctonia is the main cause for this disease, where infected plant stems show rotting just above the soil level and after some time, crop growth gives a retarded appearance. Avoid watering in mid-day during warm weather and prevent fluctuations in temperature to control *Rhizoctonia.*

Wilt

Alstroemeria wilt caused by *Fusarium oxysporum* in India was first time reported by Shanmugam *et al.,* 2007. Symptoms exhibited leaf chlorosis and slight vein clearing on outer leaflets, followed by leaf yellowing and abscission, discoloration of stem vascular tissue and death. *Fusarium oxysporum* causing vascular wilt or basal rot inflict heavy crop damage, leading to deterioration in quality and quantity of the marketable blooms and planting material. Seven hybrid cultivars of *Alstroemeria* (Alladin, Amor, Butterscotch, Capri, Pluto, Rosita and Tiara) were screened for resistance to *Fusarium oxysporum.* It was found that cv. 'Tiara' is more resistant to *Fusarium oxysporum* than other cultivar of *Alstroemeria* in terms of mean percent disease index (Shanmugam *et al.,* 2011).

Viral Diseases

It is propagated by division of rhizomes and once the virus infected material is propagated then the virus can transfer from one generation to the next. There are about ten viruses which infect *Alstroemeria* plants viz. *Alstroemeria mosaic potyvirus*, *Alstroemeria streak potyvirus, Alstroemeria carlavirus*, *Cucumber mosaic cucumovirus, Tomato spotted wilt tospovirus/ Impatiens necrotic spot tospovirus*, *Alstroemeria ilarvirus*, *Tobacco rattle tobravirus, Arabis mosaic nepovirus,Freesia cucumber potyvirus* and *Rhabdovirus*. Potyvirus infection can be easily detected through serological (ELISA, ISEM, RIA, Immunodiffusion and Immunoblottiing) and nucleic acid based techniques.Tissue culture accompanied by chemotherapy and thermotherapy are useful for quality planting material from *Alstroemeria* infected with potyvirus. Besides these, control of aphids, thrips, weeds, nematode etc. and use of resistant cultivars can also be used for controlling potyviruses (Mehra *et al.*, 2005).

Cucumber mosaic virus

Infection of *Alstroemeria* hybrids in India with a subgroup 1 strain of *Cucumber mosaic virus* (CMV) is reported for the first time in India. The virus was identified on the basis of host range, transmission by *Aphis gossypi* and *Myzus persicae* in the non-persistent manner, ELISA, electron microscopy and RT-PCR using CMV-specific primers. CMV was detected in nine hybrids and 61% of plants by dot-blot hybridization (Verma *et al.*, 2005).

Insects–Pests, Nematodes

The main insects attacking *Alstroemeria* plants are aphids, caterpillars, grasshoppers and whiteflies.

Aphids

The aphids are usually found on young leaves and flower buds. It sucks the sap of foliage and bud and causes retarded growth of the plant with poor quality flowers. It also acts as a vector for viral diseases. Two spray of malathion or rogor @ 1.0-1.5 ml per l of water at 15-20 days interval can control it.

Green Caterpillars

The green caterpillars are particularly active in summer. They roll the leaf and flower bud and damage the foliage and flowers. It can be controlled by spray of malathion or metasystox @ 1.0-1.5 ml per l of water.

Grasshoppers

Grasshoppers are particularly active in summer and rainy season. They generally attack leaves and start piercing and sucking the foliage causing the growth to become retarded. These are mostly found in closed enclosures like polyhouses. Regular spray of monocil or metasystox @ 1.0-1.5 ml per l of water on the plants and surrounding areas can control the grasshoppers.

White fly

Whiteflies mostly found on young foliage and underside of the leaves. They suck the sap of foliage and bud and causes retarded growth of the plant with poor quality flowers. They also act as a vector for viral diseases. It spreads very fast if not checked early. Regular spray of acetamiprid @ 3 gm per 10 l of water on the plant and beneath the surface of leaves can control the whitefly attack.

Nematodes

Nematode spp. *Pratylenchus penetrans* and *Pratylenchus bolivianus* generally attack the plants. It can cause loss of production and quality by affecting the vigour and growth of plants, fading of flower colour and mortality of plant. Infected plant shows light brown strains on the underground stem. Control measures include sterilization of soil by steam or drenching by formaldehyde @ 2% before planting, incorporation of Furadan in the soil. Application of neem based product and crop rotation with marigold can also minimize the nematode infestations.

Physiological Disorder

Alstroemerias are sensitive to fluoride. Irrigation water containing fluoride can cause leaf scorch. Therefore fluoride containing water and fertilizers such as superphosphate should be avoided. Leaf scorch can also be induced by widely fluctuating polyhouse temperature, which needs to be maintained properly. Flower bud abortion / blasting can be caused by very low light intensity.

Summary

It is high yielding cut flower crop with long vase life of cut flowers. Large number of grower will be benefited and could grow better quality flowers for domestic as well as export flower market which will improve considerably the economic condition of the flower growers.

An investigation was carried out to study the comparative economic analysis between cultivation of *Alstroemeria* cut flower crop and traditional crop rotation (Rice + Wheat) in 300m^2 and 1ha area, respectively in Himachal Pradesh. After three years, data were analysed and it was observed that the return of net income of Rs 0.439/m^2/year was obtained by cultivation of traditional crop rotation. For cultivation of *Alstroemeria* it was observed that since it requires high initial capital cost, therefore, net income return is negative by sale of its flowers in first and second year of crop. The maximum net return (Rs 549/m^2) was observed in second year of crop when growers become capable for selling the newly produced rhizomes/plant. By cultivation of *Alstroemeria*, it was observed that net return of Rs 128.70/m^2 may be obtained by sale of flowers in third year. If grower gets total return in third year from the capital investment made in first year, then in the fourth year net income obtained by the sale of flowers becomes three times that of the third year (Singh *et al.*, 2005).

References

Bayer, E. 1989. Die gattung Alstroemeria in chile. In: Mitteilungen der Botnischen Staatssammlung Munchen, Band 24. Botnische Staatssammlung Munchen, Munchen. p 3.

Bridgen, M.P.1997. Alstroemeria, pp. 201-209. In: The Physiology of Flowering bulbs, A. De Hertogh, M. Le Nard, editors. Elsevier, Amsterdam.

Dhiman, M.R. 2009. Effect of GA_3 and benzylaminopurine on foliar chlorosis of cut Alstroemeria stems. National conference on Floriculture for Livelihood and Profitability, IARI, New Delhi, pp2.3.

Gupta, R., and Y.C. Gupta 2009. Effect of low temperature storage and thidiazuron application on establishment of plant and shoot production in Alstroemeria. National conference on Floriculture for Livelihood and Profitability, IARI, New Delhi.16-19 March, 2009.pp2.2.

Hang, A. and T.Tsuchiya1988. Chromosome studies in the genus Alstroemeria. II. Chromosome constitutions of eleven additional cultivars. Plant Breeding, 100:273-279.

Healy, W.E. and H.F.Wilkins 1985. Alstroemeria, pp. 415-424. In: Handbook of Flowering, vol. I, A.H.Halevy, editor. CRC Press, Boca Raton, Florida.

Healy, W.E. and H.F.Wilkins 1991. Alstroemeria: The Cut flower Crop. In: V. Ball (ed.). Ball RedBook, 15th ed.Geo. J. Ball Publishing, USA. P.311-316.

Heins, R.D.and H.F.Wilkins 1979. Effect of soil temperature and photoperiod on vegetative and reproductive growth of Alstroemeria 'Regina'. Journal of the American Society for Horticultural Science. 104:359-365.

Hofreiter, A. and E.F. Rodriguez 2006. The Alstroemeriaceae in Peru and neighbouring areas. Revista Peruana Biologia, 13:5-69.

Hoshino,Y., Murata,N. and K. Shinoda 2006. Isolation of individual egg cells and zygotes in Alstroemeria followed by manual selection with a microcapillary-connected micropump. Annals of Botany, 97:1139-1144.

Huxley, A., Griffiths, M. and M. Levy 1992. Alstroemeria. In: The New Royal Horticultural Society Dictionary of Gardening, vol.1. Stockton Press, New York.p141.

Manikrao, D. R.2007. Studies on possible alternatives of silver thiosulphate in the postharvest handling of Alstroemeria and Carnation cut flowers. Ph.D. Thesis, Dr. Y.S. Parmar University of Horticulture and Forestry, Nauni, Solan, H.P.

Mehra, A., Chandel, V., Singh, M., Verma, N., Singh, M. K, and A.A. Zaidi 2005. Viruses of Alstroemeria: Investigation, Control and Production of Virus free Plants. Bhartiya Vaigyanik Evam Audyogik Anusandhan Patrika, 13(1):30-37.

Singh, M.K. 2005. Production technology of Alstroemeria as a Cut Flower –A Review. Bhartiya Vaigyanik Evam Audyogik Anusandhan Patrika, 12(1):34-38.

Singh, M.K. and R.S. Dadwal 2005. Comparative Economic Analysis between Cultivation of Alstroemeria as a Cut Flower in Polyhouse and Traditional Crop Rotation. Bhartiya Vaigyanik Evam Audyogik Anusandhan Patrika, 13(2):149-153.

Singh, M.K. 2006a. Performance of Alstroemeria cultivars under polyhouse conditions. Indian Journal of Horticulture, 63(2): 195-198.

Singh, M.K., Ram, R. and R.Prasad 2006b. Effect of thinning of vegetative shoots on cut flower production of 'Serena' Alstroemeria hybrids under polyhouse conditions. The Indian Journal of Agricultural Sciences, 76(3): 181-182.

Singh, A., Nazki, I.T., Qadri, Z.A., Ahmed. Z. and Sofi Inam-ul-Rehman 2010a. Effect of storage root number and nitrogen fertigation on clonal multiplication of Alstroemeria cv 'Pluto'. National Symposium on lifestyle floriculture: Challenges and Opportunities. Dr Y.S. Parmar University of Horticulture and Forestry, Nauni, Solan, (H.P.) March19-21, 2010 pp-3.12.

Singh, A., Nazki, I.T., Qadri, Z.A., Ahmed. Z.,and Sofi Inam-ul-Rehman 2010b. Effect of storage root number and benzyladenine on clonal multiplication of *Alstroemeria* cv 'Pluto'. National Symposium on lifestyle floriculture: Challenges and Opportunities. Dr Y.S. Parmar University of Horticulture and Forestry, Nauni, Solan, (H.P.) March19-21, 2010. pp-3.12.

Shanmugam,V., Kumar, S., Singh, M.K., Verma, R., Sharma,V., and N.S.Ajit 2007. First report of Alstroemeria wilt caused by Fusarium oxysporum in India. Plant Pathology, p.1587.

Shanmugam,V., Atri, K. and M.K. Singh 2011. Screening Alstroemeria hybrids for resistance to Fusarium oxysporum. Indian Phytopathology, 64(4):388-389.

Stephens, J.L., Tsuchiya, and T. H. Hughes 1993. Chromosome studies in Alstroemeria pelegrina L. International Journal of Plant Sciences, 154:565-571.

Srasburger, E. 1882. Über den Teilunsvorgang der Zellkeme und das Verhältnis der Kemteilung zur Zeleteilung. Archiv für Mikroskopische Antomie, 21:476-590.

Taylor, W.R. 1926. Chromosome morphology in Fritillaria, *Alstroemeria*, Silphium and other genera. American Journal of Botany, 13:179-193.

Traub, H.P. 1943. Dichogomy and interspecific sterility. Hebertia, 10:131-132.

Van Scheepen, J. (Editor), 1991. International Checklist for Hyacinths and Miscellaneous Bulbs. Royal General Bulb Grower's Association (KAVB), Hillegom, The Netherland, pp409.

Verma, N., Songh, A.K., Singh, L., Raikhy, G., Kulshrestha, S., Singh, M.K., Hallan, V., Rajaram and A.A. Zaidi 2005. Cucumber mosaic virus (CMV) infecting Alstroemeria hybrids in India. Australasian Plant Pathology, 34:119-120.

Wazir, J.S., Sharma, Y.D. and S.R. Dhiman 2010. Response of potted Alstroemeria to different growth retardants' under protected conditions. National Symposium on Lifestyle Floriculture: Challenges and Opportunities. Dr Y.S. Parmar University of Horticulture and Forestry, Nauni, Solan, (H.P.) pp-5.9.

7

Hi-tech Cultivation of Lilium Under Protected Cultivation

M. Kannan, S. Vinodh and P. Ranchana

Department of Floriculture & Landscaping, Horticultural College & Research Institute, Tamil Nadu Agricultural University, PN Pudur, Coimbatore Tamil Nadu – 641003

Introduction

Flowers speak millions of unspoken words. Presenting a bouquet and making arrangements with flowers in most of the drawing rooms, offices, hotels and hospitals have become the style of the day in the fast – moving modern world where there is hardly any time for exchange of sweet memories. Since time immemorial, the importance of such a delicate creation of nature has always been appreciated. Tamil Nadu ranks first among the flower producing states of India, contributing 25% of the country's flower production with an area of 31, 970 ha under flowers. Of this, around 700 ha are under protected cut flower cultivation. In Nilgiri district alone, it has an area of around 70 ha under protected cut flower production which is inclusive of around 11.2 ha of lilium.

The genus Lilium (*Lilium* spp.) is a herbaceous flowering plant normally grown from bulbs, about 110 species world over belongs to lily family Liliaceae. It is native of the northern temperate regions. The species in this genus are the true lilies while the other plants with lily in the common name are related to the other groups of plants. Lilies are usually erect leafy stemmed herbs. The majority of species form tunicless scaly underground bulbs from which they produce flowers. The large flowers have three petals along with three sepals, look like petals.

Types of Lilium

The five main groups of lilies are Asiatic, Oriental, OT, LO and LA hybrid. Although Asiatic and Oriental lilies are well known to the general public, there have been many new releases of LO, LA, and OT hybrid lilies brought onto the market. These new hybrids provide an opportunity to introduce to the public the new possibilities in stem length, flower shape, size, colour, and scent. Growers can expect to see many more new and Hybrid lilies which are the result of genetic crossings between lily species, making newer lily cultivars difficult to classify into conventional groupings. The OT hybrids are a cross between Oriental and Trumpet lilies. Oriental and OT lilies are producing fragrant flowers with some cultivars having stronger scented flowers than others. For customers sensitive to strong odours, LA hybrids may be a better choice. The LA hybrids are a cross between *L. longiflorum* and Asiatic lily. The LA hybrids are increasing in popularity compared to the Asiatic hybrids. LA hybrids generally have a larger flower size and the flowers are more clustered at the top of the stem and are more upward facing than Oriental lilies. The LO hybrids are a cross between *L. longiflorum* and Oriental lily. LO hybrids are the newest addition of lilies and will provide unique flower shapes, sizes and colours.

Importance of Lilium

Lilies are very useful flowering plants which can be planted in any locations as they do well in bed or border with others perennial. Almost all varieties of lily are suitable for planting in combination with other plants and in front of tall shrubs. They look very charming in large clusters for mass effect. Shorter varieties are planted in containers. Lilies may be planted in the perennial border, preferably with the plants having ample foliage.

Commercial Hybrids

1. Asiatic Hybrid Lilies

Colour	Cultivar	Bud count	Stem Length (cm)
Deep Red	Black Out	3–6	85
Red	Monte Negro	3–5	75
Orange	Brunello	4–6	85
	Lyon	3–5	80
Orange/Maroon	Loreto	2–5	75
Pink	Vivaldi	3–5	85
	Toronto	5–7	90
Salmon	Cannes	3–5	95
	Farfalla	4–7	90
White	Navona	4–7	85
	Sorpressa	2–4	90
Pink/White	Vermeer	2–4	90
White/Pink	Renoir	4–7	90
Pink/Yellow	Corrida	4–6	85
Yellow	London	4–6	75
	Pollyanna	4–7	90

2. LA Hybrid Lilies

Colour	Cultivar	Bud count	Stem Length (cm)
Pink	Algarve	4–6	110
	Brindisi	3–5	90
Yellow	Dazzle	5–8	100
	Glow	5–8	90
Red	Fangio	3–6	115
	Original Love	3–5	105
White	Litouwen	4–6	110
	Timaru	3–5	100
Salmon	Menorca	3–6	105
	Salmon	3–6	80
	Classic	4–6	90
Orange	Madrid	3–6	110
	Royal Trinity	3–5	85
White/Pink	Samur	4–6	95
Pink/Yellow	Royal Sunset	3–5	95
Lime Green	Courier	4–6	110

3. Oriental Hybrid Lilies

Colour	Cultivar	Bud count	Stem Length (cm)
Dark Pink	Acapulco	4–6	110
	Mero Star	3–5	90
	Tiber	5–8	100
	Star Gazer	5–8	90
Dark Red	Dordogne	3–6	115
	Tropical	3–5	105
Pink/Yellow/ Red	Bergamo	4–6	110
White/Yellow	Legend	3–5	100
White/Yellow/Pink	Maryland	3–6	105
Pink	Lombardia	3–6	80
	Sorbanne	4–6	90
White/Pink	Marco Polo	3–6	110
	Muscadet	3–5	85
White	Casa Blanca	4–6	95
	Siberia	3–5	95

4. OT Hybrid Lilies

Colour	Cultivar	Bud count	Stem Length (cm)
Yellow	Conca d'Or	4–6	100
	Yelloween	6–8	130
Pink	Gluhwein	4–9	120
	Maywood	5+	110
Yellow/Red	Nymph	5+	110
Red /Yellow	Shocking	4-6	120
Red	Visaversa	6-9	110

5. LO Hybrid Lilies.

Colour	Cultivar	Bud count	Stem Length (cm)
White/Raspberry	Triumphator	5-7	100
White	White Heaven	5-7	105
	White Elegance	5-7	100

Countries/Companies Importing Bulbs

The commercial production of bulbs of Lilium requires a cool environment, which is not found under Indian conditions. For this reason, most bulbs used for cut-flower production in India are produced in Netherlands, Chile, New Zealand, South Africa, France and the northwestern United States in Oregon and Washington State. Companies like, Van den Bos BV, VWS – flower bulbs, Royal Van Zanten and Onings BV are the major bulb importers to India.

Points should be considered while procuring bulbs

- Bulb size, colour, bud count, flower position and time of procurement
- Cultivars should have bud count of 3-5, upright facing, sturdy stem
- Asiatic bulb size should not be less than 10 -12 cm & for Orientals lily – 14-16 cm
- Free from pest & diseases
- Should reach the polyhouse/planting site without damage

Bulb Storage and Handling

Lily bulbs ready for production are available year round from the bulb suppliers. They pre-cool the bulbs for 6 to 8 weeks at 34 to 36°F (1.1°Cto 2.2°C) to ensure that the bulbs produce flower more evenly. Once the bulbs are precooled, suppliers freeze them in peat moss at 28 to 29°F (-2.22°C to -1.66°C) for storage. Freezing the bulbs prevents sprouting, reduces loss of bulb energy reserves and minimizes disease occurrence.

Improved Production Technologies

There are several production methods available when growing lilies as a cut flower crop. Lilies can be grown in raised beds in the field, in high tunnels, and in greenhouses. Lilies can also be grown in crates in greenhouses. Before choosing a production method, or a combination of methods, growers should consider factors such as facilities and resources available, the types of lilies to be grown, and the time of year for growing the lily crop. Fall and early spring production of lilies requires the use of high tunnels.

Site Selection and Management

Select a site that receives a minimum of 6 to 8 hours of sun daily to allow the flowers and foliage to dry off by evening, which decreases the chance of Botrytis. Choose a loam soil with 2 to 5 percent organic materials that is at least 6 to 8 inches deep. The soil should be well drained and have a pH of 6.3–6.8. Apply a pre-emergent herbicide soon after planting.

Greenhouse Crate Production

Lily bulb shipping crates can be reused for production in the greenhouse. Place newspaper on the bottom of the crates to prevent losing substrate. Use a well-drained commercially available soilless substrate. Your own mixture of leaf compost, coarse peat, and perlite or sand in 1:1:1 ratio could also be used.

Place a 2-inch layer of substrate at the bottom of the crate, lay the bulbs out, fill the rest of the crate with substrate, and water well. Keep the crate in a cool area between 50 and 55°F (10 – 12.5°C) for 7 to 20 days before placing it in the greenhouse to keep the bulbs from developing shoots before establishing an adequate root system. It is important to take the trays out of the cool area and into the light, whenever sprouts have developed 2 to 3 inches in height.

Treatment of Bulbs

Dip bulbs for about 20 minutes in a solution of Emisan (0.2%), Thiram (0.3%), Captan (0.2%), Bavistin (0.2%) or Benlate (0.2%). Dry in shade before planting or storing. Before planting treat bulbs in systemic fungicide and before storing in contact fungicide. These must be thoroughly dried before planting or storage.

Raised beds

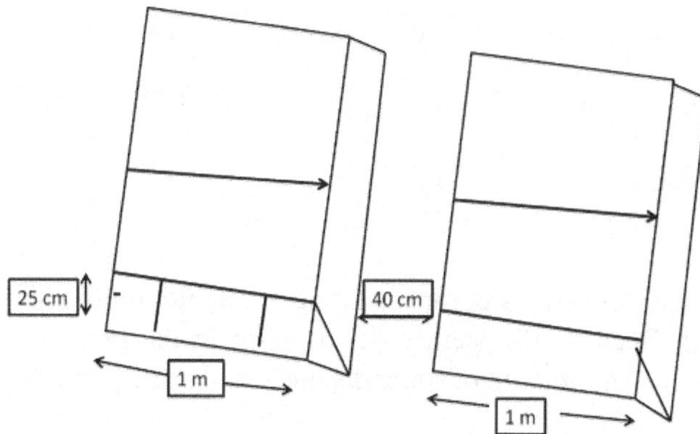

Lilium bulbs are planted at 6 inch depth with a spacing of 15 x 10 cm. Based on the bulb size, number of bulbs per row will vary slightly. After planting, the beds are prepared with convenient length, 1 m width and 25 cm height from the ground level. Between the two beds a spacing of 40 cm is left for walking space. Planting density depends on cultivar, bulb size and time of the year, with a range of 25-60 bulbs/m^2

Light Requirement

Light is very important factor for lily culture. High light intensity in summer reduces the stem length and therefore 50% shade nets are recommended to cover the crop. Low light intensity in winter leads to flower abortion and abscission. Supplementary lighting during winter increases yield, stem sturdiness and quality of flowers.

Supplemental Lighting

Lilies are given supplemental light from late September until early April to extend the day length to 16 hours. The number of hours of lighting necessary to manipulate the day length to 16 hours will vary depending upon the exact time of year. Increased light levels will also help to avoid bud abortion. Start lighting plants when the foliage emerges. Initially you can use 2,500 watts of halogen lighting or one high intensity discharge (HID) light per 100 sq. feet placed 6 to 8 ft above the crop. The use of HID lights is preferable to incandescent or fluorescent lamps because of high efficiency, uniform light distribution, and a lower amount of shading.

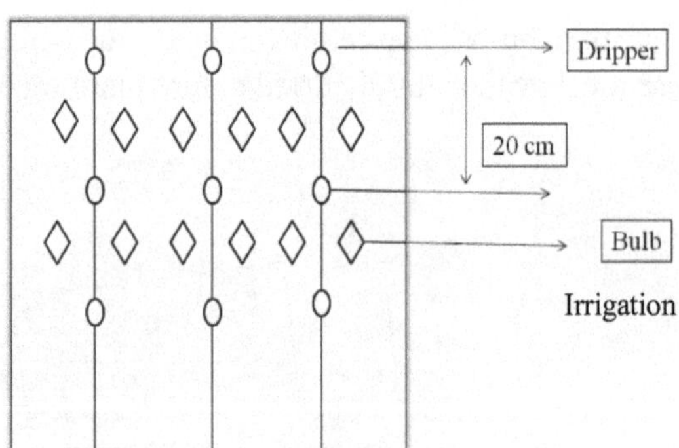

Moisten the soil a few days before planting to enable rooting to start straight after planting. Because the stem roots develop in the top layer, it must be constantly moist. The amount of water depends on type soil, greenhouse climate and the variety.

Water requirement in summer : 6 – 8 lit/m^2/day

Water requirement in winter : 5 – 6 lit/m^2/day

For first two weeks, irrigate only by using water can or shower. From third week onwards it is recommended to use drip for irrigation.

Fertigation

Sustainability of any system requires optimal utilization of resources such as water, fertilizer or soil. Micro irrigation in the form of drip irrigation has led to revolutionary changes in crop production, resource conservation and resource use efficiency in many crops. There is considerable saving in water by adopting this method since water could be applied almost precisely to the root zone. By introducing fertigation, it is possible to increase the yield potential by three times with the same quantity of water, by saving about 45 to 50 per cent of irrigation water and increasing the productivity by about 40 per cent. When fertilizer is applied through drip irrigation, it was observed that the yield has been increased and about 30 per cent of the fertilizer could be saved.

Proper fertilization of lilies used as cut flowers is important. The soil pH should be in the 6.3 to 6.8 range for optimum nutrient uptake and plant growth. This pH range is good for growing lilies both in the soil media and in soilless substrate in the greenhouse. For greenhouse production of lilies, soil samples should be taken every week during the growing season and the soluble salt and pH monitored using a portable pH and EC meter. Acceptable EC for production of Lilium flowers should be between 1.0 and 1.5 ms/cm. High soluble salt levels can cause soft stems, leaf burn, and reduced plant height because of inadequate root development.

Fertigation Schedule

- Recommended Dose: 4:2:5 kg NPK/ $500m^2$
- 100% of N&K& 25% of P applied through fertigation @ 4:0.5:5 kg NPK/ $500m^2$
- Water soluble fertilizers such as Polyfeed (19:19:19), MAP (12:61:0), KNO_3 (13:0:45)and CAN (20:0:0) were applied at weekly intervals
- 75% RDF of Phosphorous is applied as Super phosphate @ 9.4 kg/ $500m^2$ as basal
- Polyfeed (19:19:19) is applied @ 2.65 kg/ $500m^2$
- KNO_3 (13-0-45) is applied @ 10 kg/ $500m^2$
- CAN (20:0:0) is applied @ 11 kg/ $500m^2$

The above fertilizers are to be applied through irrigation water at weekly intervals as per the schedule furnished below;

Fertigation Sertigation Schedule for Asiatic Hybrid Lilium

	Recommended Dose (100% RDF) - 4:2:5 kg NPK/ 500m²		4:0.5:5 kg NPK/ 500m² (75% P as soil application in the form of SSP)						
S.No.	Stages & Weeks	Fertilizers	Total nutrients (kg / 500m²)	N	P	K	% Nutrient requirements		
1.	Bulb Planting to establishment	19:19:19	0.525	0.100	0.100	0.100			
	(3 weeks)	Multi K (13-0-45)	0.890	0.115	-	0.400	10	20	10
		CAN (20-0-0)	0.925	0.185	-				
	a) Sub total		2.340	0.400	0.100	0.500			
2.	Establishment to shoot formation	19:19:19	1.050	0.200	0.200	0.200			
	(4 weeks)	Multi K (13-0-45)	4.000	0.520	-	1.800	40	40	40
		CAN (20-0-0)	4.400	0.880	-				
	b) Sub total		9.450	32.00	0.200	2.000	-		
3.	Shoot formation to flower bud	19:19:19	0.790	0.150	0.150	0.150			
	initiation (2 weeks)	Multi K (13-0-45)	3.000	0.390	-	1.350	30	30	30
		CAN (20-0-0)	3.300	0.660	-	-			
	c) Sub total	7.090	1.200	0.150	1.500				
4.	Flower bud initiation to flower bud	19:19:19	0.265	0.050	0.050	0.050			
	development & upto harvest	Multi K (13-0-45)	2.110	0.275	-	0.950	20	10	20
	(4 weeks)	CAN (20-0-0)	2.375	0.475	-	-			
	d) Sub total		4.750	0.800	0.050	1.000			
		Total (a + b+ c+ d)		4	0.5	5	100	100	100

19:19:19 – 2.65 kg/ 500m²; Multi K (13-0-45) – 10 kg/ 500m²; CAN (20-0-0) – 11 kg/ 500m²; SSP – 9.4 kg/ 500m²

Intercultivation

Hand weeding was done depending upon the intensity of weed growth.

Staking

Wire mesh or plastic nets having inner size of 15-20 cm squares are placed on the ground in three layers, which are erected at 20, 35 and 50 cm above the ground level with the growing plants. String or rope is erected in three rows at the same distance along the rows.

Nutritional Deficiency

Calcium Deficiency

Calcium deficiency is occasionally a problem in greenhouse production. Some Oriental lilies are susceptible to a condition known as "upper leaf necrosis" (ULN), which causes distortion and necrosis of the upper leaves. Problems with ULN have been reported primarily on 'Star Gazer' lilies, which have a low calcium content in the shoot and bud scales of the bulb.

The bulbs have enough calcium to supply the lower leaves of the plant, but the rest must come from the soil. Growers can improve nutrient uptake by monitoring for root rots and soluble salts. Calcium deficiencies can develop even when there is an adequate amount of calcium in the soil. In those instances, adding calcium to the soil would not be beneficial because the deficiency is a result of uneven distribution within the plant.

Calcium is taken up by the roots and translocated only to transpiring tissues. Calcium will not move into the young leaves of the plant if they are not actively transpiring. A low transpiration rate due to high humidity or low light intensity increases the likelihood of lilies developing ULN. Transpiration rates can be raised by increasing light intensity and decreasing humidity.

Avoid growing plants under shady conditions and provide supplemental lighting during the winter months. Keep the leaves and the greenhouse as dry as possible. Overlapping leaves can also restrict transpiration. On Oriental lilies, the leaves associated with the flowers are overlapped by older leaves before the buds develop. These are the same upper leaves that display symptoms of ULN.

Symptoms of upper leaf necrosis appear long after the actual damage has occurred, so most growers catch the problem after it is too late. Some preventative measures include using a calcium nitrate fertilizer and applying foliar sprays of calcium chloride. Use caution; foliar sprays can be phytotoxic if applied improperly.

Iron

Symptoms: The leaf tissue between the veins of young leaves becomes yellowish-green, particularly in plants with rapid growth. The greater the iron deficiency the more yellow the leaves look.

Control

- Soil should be well drained with low pH level
- Chelated- Fe should be applied @ 2-3 gm/m^2 before planting and maximum 2 gm / m^2 after planting.

Nitrogen

Symptoms: The whole leaf becomes lighter in colour and this is often more noticeable when plants are about to bloom. The plant often seems rather light green in appearance. Soil with a low nitrogen level produces a crop with stems which are lighter in weight with less number of flower buds. The foliage in the vase will turn yellow more quickly.

Control

- Always apply sufficient quantity of nitrogen, preferably based on the results of soil sample.
- If the nitrogen deficiency is diagnosed during cultivation apply an additional rapid action nitrogen fertilizer. However, bear the risk of leaf scorch in mind during this procedure and make sure the crop is always washed off thoroughly

Abnormalities

Leaf Scorch

Leaf scorch occurs when there is a disturbance in the balance between absorption and evaporation of water. This is the result of inadequate absorption or evaporation which causes a calcium deficiency in the cells of the youngest leaves. Cells are destroyed and eventually die. A sudden change in the relative humidity inside the greenhouse also affects this physiological imbalance, that leads to poor root system and high salt level in the soil. Large bulbs are more susceptible than smaller ones.

Management: Disease and pest which could damage the roots should be controlled effectively. Soil should be moistened before planting. It is better not to use susceptible varieties but if this cannot be avoided do not use larger bulbs as these are extra sensitive. Plant bulbs with a good root system. Plant to an adequate depth i.e. allow 6-10 cm of soil on top of the bulb. Prevent large differences in greenhouse temperature and air humidity levels during period of increased susceptibility. Try to maintain RH level of approx. 75%. Ensure that plants maintain even transpiration and avoid excess transpiration by shading. Rapid growth must be prevented

Harvesting and Yield

Asiatic hybrids take 8-10 weeks, Orientals 14-16 weeks for harvesting of flowers from planting. The spikes are to be harvested when the lower first bud turns from green to original colour of the variety, but has not yet opened. The spikes are to be cut 8 -10 cm above the ground level. After harvesting, the cut stems are to be graded as per the number of buds per stem, length and firmness of the stem. After grading, the leaves have to be removed upto 10 cm from the bottom of stem before bunching and subsequently sleeved with polythene sheets. A single bunch contains 10 stems. It is then packed in perforated cardboard box (100 x 60 x 40 cm) for transportation.

Post Harvest Handling

Place cut lily stems in water in a cooler condition at 35 to 41°F an hour after harvesting to avoid shocking the flowers. Handle lily blooms carefully at all times because the flowers bruise easily. Remove the bottom one-third of the foliage and grade the stems by size. Depending upon the market, grading can be done by the stem length or by the number of flowers per stem. Lilies may be bunched into desired quantities and sleeved, again, depending upon the market. Cut lily stems should be transported in water to prolong the vase life. Freshly harvested lilies have a vase life of 9 to 14 days, depending upon the cultivar and the environment. The vase life of freshly harvested lilies can be extended by the final customer if they are re-cut and placed in a solution containing a floral preservative. Floral preservatives can also be added to the water used for storage prior to shipping, upon receipt at the retail level, and in floral arrangements. Hybrid lilies are sensitive to ethylene; keep them away from ripening fruit, maturing foliage and flowers, or any other ethylene source. Anti-ethylene treatments are effective on many lily cultivars, especially on Asiatic lilies. Immediately after harvest, the lower portion of the cut spikes should be immersed in water for prolonging the vase life of spikes. A holding solution consisting of silver nitrate (50 ppm) + 3% sucrose is best to extend vase life, delaying leaf senescence and enhance ppost harvest keeping quality of Lilium cut flowers.

Vase life

Asiatic hybrid : 7 – 14 days

Oriental hybrid : 10 – 15 days

L. longiflorum : 8 – 10 days

Grading & Price fixation in Lilium

Grading is done based on the number of buds per stem

Grades	Average price per stem (Rs.)	
	Asiatic	Oriental
A+ - 4+ buds	30	40
A - 4 buds	25	35
B - 3 buds	20	25
C - 2 buds	15	20

Procedure for post harvest treatment of the bulbs

1. Reduce the frequency of irrigation water. Maintain soil moisture level in such a way that bulb scales should not dry out. Excessive moisture may lead to rotting of bulbs.
2. Allow bulbs to remain in the beds for 4 to 5 weeks (above ground stem portion should dry out and can be pulled out from bulb easily).
3. After 5 weeks remove the bulbs from soil along with dried stem.
4. Remove dried stem carefully without damaging the bulb.
5. Wash bulbs with clean water and treat with 2% Bavistin solution for 10 minutes.
6. Remove the bulb from solution and air dry in shade. Too much drying may lead to loosen root skin. Such bulbs; after planting may develop root rot.
7. Immediately after air drying pack the bulbs in plastic crates with moist coco peat wrapped with perforated plastic sleeves.
8. Coco peat used for packing must be sterilized.
9. Keep the crates in cold storage at 20C for 2 weeks and then at -10C for 6 weeks.
10. Keep crates open for one day in cold storage and then close with plastic sleeves.

Plant Protection

Diseases

Bulb and Scale Rot

This disease is caused by the fungal pathogens. Plants affected by these diseases are retarded in growth and the leaves have a pale green colour. The underground part of the stem may show orange-brown to dark brown stains, which afterwards become larger and spread to the inside of the stem. The infected bulb's scales will show dark brown stains and the rotting starts at the base of the bulbs and scales. The plant finally dies prematurely.

To prevent the diseases, bulbs should be planted in pre-sterilised soils. Bulbs should be dipped for one hour in 0.2% captan + 0.2% benlate to minimise the disease infection. Also keep the soil temperature as low as possible during the entire period of cultivation by frequently irrigating the field.

Foot rot

This is caused by the Phytophthora fungus. The infected plants have violet – brown spots spreading upwards. The plants are retarded in growth or may wither suddenly. The leaves start turning yellow from the bottom of the stem.

To control the diseases, sterilize the soil before planting the bulbs. Dithane M-45 may be applied @ 200g / 100 m2 as soils drench.

Root rot

This disease is caused by the fungus *Pythium*. These fungi prefer moist condition and thrive best at 250 – 300 C. The infected bulbs and stem roots show light brown spots and signs of rotting. The infected plants remain short in height leaves are narrow and dull in colour. Such plants will show more bud drop than normal plants. The flowers are smaller in size and often do not open fully.

The soil should be disinfected chemically. The affected plants may be sprayed with Dithane M-45 @ 0.2 %.

Leaf spot Disease

Leaf Spot disease is mainly caused by *Botrytis* under moist condition. Botrytis produces spores which are spread by rain and wind to nearby plants. Under dry condition the disease will not spread. When infected, the leaves show dark brown spots of 1-2 mm diameter which will increase in size to form round or oval spots. The affected leaves and flowers will ultimately die.

To control the diseases, reduce irrigation to make the soil Dry .Spray Benlate @ 5gm per 10 m^2.

Virus Diseases

Lilies are infected by different types of virus viz. Lily Symptomless Virus, Cucumber Mosaic Virus, Tulip Colour Breaking Virus etc,. The plants raised from virus infected bulbs become weak in vigour and produces inferior quality flowers. In case of severe infection, the plants become stunted and deformed. For production of quality flowers diseases free bulbs should be used.

Insects and Pests

Aphids

Aphids live only on young leaves particularly at the backside of the leaves. Young buds may also be affected resulting in deformed flowers. Spray Nuvan @ 2ml/litre of water.

Thrips

This is also a sucking type of insect. A severe attack will adversely affect the plants growth and flowering. Those flowers will not be accepted in the market. Regular spraying with Monocrotophos @ 2 ml/litre of water will protect the plants from the attack of thrips.

8

Hi-tech Production Practices in Cut Rose Under Protected Cultivation

M. Kannan, P. Ranchana and S. Vinodh

Department of Floriculture & Landscaping, Horticultural College & Research Institute, Tamil Nadu Agricultural University, Coimbatore – 641003

Introduction

Rose, acclaimed as the "Queen of Flowers", is one of the most popular cut flowers in the global floriculture trade. Popularly referred to as 'Dutch rose' in the international market, cut rose occupies a prominent place among the cut flowers owing to its exquisite form, size and attractive colours. Apart from cut flower it is used for making various value added products such as rose oil, rose water, garland, gulkhand, rose attar etc and also in landscaping as pot plant and garden plant. The present day rose (*Rosa hybrida*) combines all desirable qualities such as, productivity, attractiveness of buds, a variety of colours and forms of the typical species of *Rosa*.

The major cut rose producing countries of the world are the Netherlands, Colombia, Kenya, Israel, Italy, United States, and Japan. In India, roses are mostly grown under cover (poly house) in Nasik, Pune, Hosur, Kalimpong, Darjeeling, Bangalore, Solan, Palampur, Shimla, Srinagar, Delhi, Ludhiana and Calcutta. The roses grown for export should be popular varieties of that region and maintain high quality standards with respect to bud size and stem length and should have lush green leaves and be free of pests and diseases. These high quality standards can be achieved and maintained only when grown inside poly house or greenhouse, which gives protection against inclement weather conditions and pests and diseases.

Varieties

The following varieties are grown commercially in greenhouses for cut flower production

Flower colour	Variety
Red	First Red, Grand Gala, Taj Mahal, Happy Hour, Red Corvette, Jaguar, Gabrilla, Sasha, Grand Gala, Dallas, E.G. Hill, Happiness, Happy Days
Yellow	Gold Strike, Skyline, Golden Times, Golden Rapture, Aalsmeer Gold, Cocktail, Papillon, Texas
Pink	Kiss, Europe, Prophyta, Noblesse, Pink, Aristocrat, Better Times, Vanity, Fair, Pavaroti, Vivaldi, Flirt, Florence
Orange	Movie Star, Miracle, Tropical Amazon, Indian Puma, Candid, Mercedes, Jazz, Orange Delight, Lambarda
White	Ice Berg, Polo, Holly Wood, Avalanche, Tineke, Eskimo, White Pearl
Cream	Prestige, Vivaldi, Verselia, Florence
Bicolour	Amour, Rodeo, Confetti, Ambience, Lionides, Yellow Gloria

Varieties released from IIHR, Bangalore

Arka Parimala

Pedigree	cv. Red chief
Breeding method	Half sib selection
No. of flowers per plant per year	80-100
Colour of flower and special traits	Red, fragrant, moderately resistant to thrips and black spot
Utility	Loose flower and as short stalk cut flower

Dr. G.S. Randhawa

Parentage	Queen Elizabeth x American Heritage
Breeding method	Hybridization and Selection
No. of flowers per plant per year	70
Colour of flower and special traits	Peach pink colour fading to salmon pink along the border. Floriferous with pointed buds
Utility	Garden display

Kiran

Parentage	Big Red x Blue Moon
Breeding method	Hybridization and Selection
No. of flowers per plant per year	40
Colour of flower and special traits	Flowers tyrian purple fading to rose bengal with white streaks on some petals
Utility	Garden display

Nishkant

Parentage	*Rosa multiflora*
Breeding method	Mutation-Isolation of bud sport
Growth habit	Vigorous, bushy and no thorns
Utility	Used as root stock only

Varieties Released from IARI, New Delhi

Varieties of HT roses are Abhisarika, Anurag, Arjun, Charugandha, Chitwan, Dr. B. P. Pal, Dr. Bharat Ram, Ganga, Jawahar, Mother Teressa, Mridula, Mrinalini, Nurjahan, Priyadarshini, Pusa Sonia, Pusa Gaurav, Pusa Bahadur, Pusa Priya, Pusa Garima, Raj Kumari, Raktagandha, Rangasala, Srabhi, Vasant.

Varieties Released from NBRI, Lucknow

Varieties of HT roses released by NBRI, Lucknow are Light Pink Prize, Mrinalini Stripe, Pink Montezuma, Summer Holiday Mutant, Winter Holiday Mutant.

Propagation

Budding

It is the most popular method of propagating cut roses. Shield or 'T' budding is the method commonly used. In this method a 'T' shaped incision is made on the stem of the rootstock. The bark portion from the top and sides of incision are raised up aside. The bud of the selected scion is inserted into the incision and then tied with suitable wrapping material. The scion bud and incision may be treated with root promoting hormones. Budding should be done at lower heights from the ground. To keep the stock-scion union wet, intermittent spraying is done. The inserted bud gives rise to a shoot. When the shoot is 10-12 cm long, the top portion of the rootstock above the scion is cut off and the wrapping material is removed. The bud branch is nourished by the rootstock and it grows vigorously. Success of budding depends on factors like age of the bud, time of budding and type of rootstock.

Selection of Budded Plants

The age of budded plants should be two months old with a height of 30 cm, possessing vigorous shoots and free from dieback disease.

Soil and Field Preparation

Well drained red sandy loam soil with organic matter content is highly preferred for rose cultivation. The soil pH 5.5- 6.5 and Ec - <1 is ideal for successful cultivation. Land should be thoroughly ploughed with disc plough at 60 cm depth for 2 to 3 times. Generally, the soil amendments like coirpith-3 kg, well decomposed farm yard manure - 15 to 20 kg, vermicompost-1 to 2 kg, phorate – 2 g, neem cake – 1 kg per m2 are applied at last time of plough to improve the texture and nutrient status of the growing medium. The organic manures should be mixed thoroughly with soil followed by pre-planting sterilization of soil with chemical fumigants so as to eradicate pathogens and pests.

Climate

Temperature

The ideal temperature range for growing cut roses is 18-27°C. Most commercial rose cultivars are best grown at 15.5°C night temperature. Lower temperature around 15° to 16°C improves flower quality. Increase in temperature (> 27°C) leads to decrease in stem length and loss of colour of flower buds. Some varieties form bull heads and blind shoots (abortion of the initiated bud) at low temperatures.

Temperature Management

To create a favourable environment for plant growth, side ventilation opening was altered depending upon the season, whenever the temperature went high, the rollable polyethylene flap was used to roll up, and sufficient irrigation through hose was given to bring down the temperature. While, under low temperature condition, the rollable polyethylene flap was rolled down in order to conserve the heat inside the polyhouse.

Humidity

The ideal humidity range is 60-65 per cent. High relative humidity is a major problem in some parts of the tropics. High humidity invites fungal infections namely, black spot (*Diplocarpon rosae*) and downy mildew (*Pernospora sparsa*). These two fungi cause serious damage to the leaves and finally lead to death of the plant. Good ventilation that permits air movement both within the greenhouse and out of the greenhouse is essential to prevent these pathogens. Air movement will lower the humidity and dry up the plants. During rainy season, greenhouse vents should not be closed.

Light

Light is an important factor which decides the growth, yield and quality. Increased light intensity increases flower production and quality. The spectrum of light available to the plants may also influence growth and flower colouration. Use of polythene films with UV filters is recommended.

Fumigation

Appication of Basamid GR as preplant soil sterilant is found to be effective for controlling controls nematodes, weeds, soil borne fungi, bacteria and insects. The steps involved are as follows.

Step – 1 : Prepare the soil to a fine tilth.

Step – 2 : Irrigate and leave for one week to activate the soil organisms (e.g. nematodes, fungi, bacteria etc.) & weed seeds.

Step – 3 : After one week, prepare beds with friable crumb structure

Step – 4 : Spread Basamid uniformly on the surface of the beds at the rate of 30g / m^2

Step – 5 : Press the bed by rolling over with heavy/loaded drum or wooden plank or using a roller

Step – 6 : Seal the soil by polythene sheet and prevent the active gases from escaping. Cover the beds with Polythene sheet and seal the margins tightly

Step – 7 : After 5 days (under normal temperature of 25°C- 30°C and no rains) loosen the soil to 15 cm depth with the implement, which is used for incorporation of Basamid.

Step – 8 : Leave open the bed for 2 - 3 days to allow the escape of toxic gases.

Step – 9 : Conduct germination test. If the germination is normal undertake planting.

Planning and calendar of operation

1	2	3	4	5	6	7
8	9	10	11	12	13	14
15	16	17	18	19	20	21
22	23	24	25	26	27	28

1-7 days : Soil preparation

8th day : Basamid application

9-14 days: Time required for Basamid action (6 days)

15-17th day : Duration for allowing toxic gases to escape (3 days)

18-21 day: Germination test (4 days)

22-28 days:Planting

Bed Preparation and Planting

Generally, the basal fertilizer dose of single super phosphate @ 390 g/m2 should be evenly spread and thoroughly mixed with the media before bed preparation.

Apart from the above fertilizer, bio-fertilizers and bio-control agents for the control of pest and diseases can be incorporated to soil at the time of bed preparation. *Azospirillum, Phosphobcteria, Trichoderma viridi, Pseudomonas fluorescens*, VAM each 1 kg can be added for 500m2 area for enriching the soil. Between the two beds a spacing of 45 cm was left for walking space. The plant spacing of 30 x 15 cm and 12 plants / m^2 were maintained.

Details of Polyhouse (Naturally Ventilated Polyhouse)

The naturally ventilated polyhouse (NVP) was oriented in East - West direction with central height of 5.7 to 6.5m. The frame was constructed with galvanized iron pipe. A rollable low density polyethylene (LDPE) flap was provided on all the sides of the polyhouse to control the ventilation area and to cover the side vents during rainy season to avoid the entry of rainwater and cooling effects inside the polyhouse. Glazing was provided with 200ì (800 gauge) thick ultra violet stabilized low density polyethylene film. The temperature (25 -300C) and relative humidity (70- 85%) inside the polyhouse were maintained by watering and over head sprinkling.

Layout of Drip System

Water was pumped through motor and it was conveyed to the main line after filtering through screen filter. From the source line, water was taken to the field through main line of 2" PVC pipes. Fertigation tank was installed for fertigation. From the main pipes, 1.5" PVC pipes were fixed as sub-main. From which two laterals of 12mm OD were taken for three replications. There were two sub-mains with tap control for imposing drip irrigation and fertigation treatments. Along the laterals, emitters with a discharge rate of 4litres per hour were fixed at a spacing of 30cm.

Design data

1. Length of each lateral from sub main (12mm OD LDPE) - 15m
2. Emitter spacing - 30 cm
3. Lateral spacing - 30 cm
4. Emitter type - Inline dripper
5. Emitter discharge rate - 4lph
6. Filter size (screen filter) - 100µ

Special Horticultural Practices

Building of young plants

First Growth

Rather soon after planting (1-3weeks, depending on the temperature) 2- 3 eyes/ branch will sprout. They will grow until flowering in 5-6 weeks. Since the plants don't have many roots to start with, this first growth will usually be no longer than 20cm. Some of this first growth can be blind shoots. These shoots must not be cut or bend, but left upright. Other shoots do form buds. When most of the buds show colour, knock of the buds and left the suckers grow as they come. The blind shoots can sprout as well.

Second Growth

After knocking the buds, second growth will come to flower again in 4-5 weeks. By then the vegetation will be 50-60cm high, which is necessary to have enough growing speed in the plants and to have enough foliage for bending. Wait until the maturity of the branches are showing big buds, nearby colouring. Then bending has to be started.

Bending

Bending is necessary for keeping enough leaves on the plants and the leaves are important for the production of sugars. This mass of leaves is often called the lungs of the plant. From each plant, a minimum of four stems, either with flowers or blind shoots, must be bent. For the blind shoots, take out the growing tips, to avoid the new growth on the top after bending. The place where to bend is as close to the original bush as possible (maximum 5 cm), without breaking the branches. To avoid breaking, it is advisable to do the bending in the afternoon and to create two 45 degree bends rather than one 90 degree bend. The bending should be such that the tops of the stems are below the horizontal. This is important for the apical dominance of the plant. When the stem is not bent below horizontal, eyes on the end of the stem will sprout. These ends are not sturdy. So these stems will sprout and will become thin and curved.

Such types of stems have to be bent in the direction of the path. One or two small branches may be bent into the middle of the bends. Make sure that the bent branches are not lying on the top of each other or on the base of other plant, for they catch away the light from the underlying branches.

Bending and Application of Growth Regulators

New vegetative shoots will be produced from pruned plants after 30 days. The well grown shoots are selected after attaining pencil thickness size for bending.

The shoots are twisted and gently bent at the shoot junction bud. Foliar application of plant growth regulator (BA 200 ppm) is to be done immediately after bending. BA @ 2000ppm is to be sprayed at fifteen days interval till flowering (2 months).

Bottom Breaks

Soon after bending, the first bottom breaks will be coming from the base of the plants. These bottom breaks are most important for the life of the plant, because they will carry the production. Let these bottom breaks come to flower and harvest them.

- on 4 to 5 leaves – for the strong ones
- on 2 to 3 five leaves – for the thinner ones

The shoots that comes from the places where the very first flowers were harvested can be harvested at 2 five leaves. Very thin stems that may come from the base of bend branches can be harvested very low or eventually be bend in case there is a place for more foliage.

Second and Third Crop

After cutting the bottom breaks in the proper way, the sprouting will start again very soon with two to three sprouts (generally). These sprouts will become flowers again approximately six to seven weeks after cutting the bottom break and will give the first real quality production. These flowers should be cut at 2 good five-leaves from below.

When third crop is arriving the blooms can be cut at 1 or 2 five-leaves, depending on the height of the shoot and the thickness.

Wild Shoots

With the development of the shoots, there will also come some wild shoots. These shoots have to be removed when minimal 5 cm long, every week. Especially, the plants from which the good eye is coming slow will develop rather much of these wild shoots. Be removing wild shoots regularly and in time, the good eye will probably never come out. However, we have to be careful not to take off the good sprout. Usually it has a different colour. Most wild shoots are pale green whereas a normal shoot is purple in the beginning and dark green later on. The wild shoots have to be removed completely. There are 32 sleeping eyes on the base of the shoots. It is very important to remove these eyes: as well otherwise they will sprout when the first shoot is removed. To remove the shoot well enough it is best to break it off. Do never cut it off, because, then some eyes may not be removed.

Pinching

Removal of unwanted vegetative/floral growth from the axil of leaf below the terminal bud is termed as 'pinching'. The rose bush is regulated to flower in peak seasons. This can be achieved by adopting cultural practices. 'Pinching' is one such practice followed first during the third or fourth week of October to produce rose flowers for the Christmas season. The same operation produces the second flush of flowers in February for Valentine's Day. Most of the commercial cultivars take about five-and-a-half to six weeks from pinching to produce flowers during summer and about eight weeks during winter.

Disbudding/ Topping

Removal of undesirable axillary buds and shoots was done after development of main bud in the growing shoot. The main objective of this operation is to keep the stem in standing position as bending operation is not to be carried out in summer months. So it is recommended to do disbudding of week stems. In Hybrid Tea roses, only one or two buds should be allowed to flower upon each shoot so as to have a large sized bloom. All other buds should be removed or disbudded.

Bud Capping /Netting

The nylon made small net like caps were covered to the developing bud to regulate the shape and increase the size of the bud and improve the quality of the flower.

Desuckering

The removal of damaged and surplus new growth in the following weeks after pruning should be followed as a routine. Any sucker appearing from the stock should be removed promptly.

Resting the Plants Through a Dry Period

Resting of plants is done during off season (June-August). Irrigation and nutrient application are withheld after the last flower is harvested. The dry period lasts for 4 to 8 weeks. Most of the leaves will drop. Then the plants are pruned down to 30-60 cm above the soil. After pruning, irrigation and feeding are started gradually. New shoots require shading from very strong sunlight. This method may give few bottom breaks and too weak stems. Another method is to harvest through under cutting which also decreases the height of plants. Some growers do not harvest the flowers but remove the flowers and buds when they have reached a certain stage. Before the export season starts, hard pruning is done which is said to give good bottom breaks and stronger stems.

Hoeing

Light hoeing is a very effective way to keep the soil porous so that light, air and water may reach the roots better to improve moisture retention capacity and to keep the beds free from weeds. Roses are shallow rooted plants and roots of established plants tend to grow near the surface. Hence, shallow hoeing is preferred, as deep cultivation destroys the rootlets. Hoeing should be done when soil is moist or dry but not wet and soggy.

Pruning or Under Cut

It should be done in the month of June – July. The selected plants were pruned for standard height at 50 cm from ground level.

Weeding

Control of weeds in rose fields is one of the most expensive operations in commercial cultivation. Hand weeding is normally taken up to keep the rose plants free from weed competition. However, hand weeding is laborious, time consuming and expensive and if not properly done, it would damage the plant and root system. The use of herbicides appears to be economical, convenient and efficient in controlling weeds in the rose field. Better control of monocot weeds with Glyphosate @ 1.0 kg a.i./ha and dicot weeds with Oxyfluorfen @ 0.5 kg a.i./ha can be achieved.

Fertilizer Application Through Drip Irrigation

Drip irrigation system has to be installed for the complete cropped area. It is to be done on alternate days @ 4lph for 2 hours. The fertilizer sources for supplying NPK through drip irrigation are Calcium nitrate (15:0:0), All 19 (19:19:19) and Sulphate of potash (0:0:50) respectively. Fertigation is to be given as per the fertigation schedule.

A. tank: 250 lit of solution/ac/day	
$CaNO_3$	5.6 Kg
KNO_3	3.2 Kg
$NH4NO_3$	2.8 Kg
Fe DTPA	280 g
B. tank: 250 lit of solution/ac/day	
$MgSO_4$	4.8 Kg
MAP	1.6 Kg
MKP	1Kg
SOP	1.6 Kg
Boron	24 g
$MnSO_4$	48g
$ZnSO_4$	9.6g

$CuSO_4$	9.6 g
Mo	4.8 g
KNO_3	2.4Kg

Application of Micronutrients and *Bacillus* spp

The chelated micronutrient mixture (contains Fe- 4.0%, Zn – 4.0 %, Mn-1.5 %, Cu-1.5 %, B- 0.5%, Mg-9.0%, and Mo-0.1%) has to be applied through foliage. Foliar application of BA 200 ppm, micronutrient mixture, soil and foliar application of *Bacillus megaterium* and *Bacillus amyloliquefaciens* are to be applied at fifteen days, ten days and seven days interval respectively.

Inter Cultivation

The entire rose beds are to be kept kept weed free by hand weeding at regular intervals. The scraping of rose beds is to be done once in a month.

Irrigation

The water requirement of rose plants depends on the soil type, humidity, temperature, day length, stage of growth (leaf area) and the radiation intensity. Generally, rose plants require water at 8 to 10 litres/m^2 a day, depending upon the season and crop growth.

Application of water through hosepipe sprinkler has to be done for plants at two days interval in summer months, to reduce the field heat and to maintain temperature and humidity inside the polyhouse.

Harvesting

Roses should be harvested at tight bud stage when one or two petals begin to unfold. Roses cut too early may also contain very little carbohydrates and may not open. Besides, due to insufficient hardening of neck region, the peduncle may not be able to support the weight of the bud, which droops in the neck region. This is referred to as 'bent neck'. Stage of harvest also varies greatly with the variety, distance to the market place, prevailing temperatures and consumer satisfaction. For local markets, buds can be harvested at more advance stage of opening. Flowers should preferably be cut during cooler hours of the day i.e. in the morning or evening. Stems should be cut leaving at least two 5-leaflet leaves on the stem. Stems should not be cut very close to the base as the basal portion of the stem is hard and absorbs very little amount of water. Cutting the stem near the base may also result in blind shoots in the subsequent season.

During harvesting operations, care should be taken not to damage the foliage and the petals. Especially some varieties like First Red have long thorns and hence, should be handled with extra care otherwise the thorns may damage the leaves and petals on the other stems and spoil the quality of bloom. After harvesting, the stems should be put in buckets containing water. If kept out of water, the air enters the stem and block xylem vessels. Too many stems should not be put in a bucket to avoid damage to the leaves by the thorns. The buckets should by cleaned daily be detergents.

Pre cooling

After harvesting, cut stems are to be kept in clean bucket with water containing 100 ppm Aluminum sulphate solution and pre-cooled for 5- 10 hrs at < 5ºC in cold room and graded based on the length of stem and size of the bud. Then the flowers should be shifted to cold rooms (3 - 5ºC). Cooling removes field heat from the flowers, puts down the respiration, lowers water loss and arrests the excessive opening of the bud. Uncooled flowers wither very rapidly. For red coloured varieties, the temperature of cold room should be kept a little higher as low temperatures cause blackening of petals in these varieties. The floor of cold room should be cleaned with bleach solution every week to keep away the pathogens. Never keep open flowers in the cold rooms as they are breeding grounds for botrytis.

Grading

After cooling, the stems are shifted to air-conditioned grading room. In the grading room, defective stems with blemishes or symptoms of diseases of pest infestation are removed. The stems are sorted out in different grades of length by colour coded board or automatic grading machines. The quality parameters for grading are :

1. Long stemmed varieties are graded from 40 cm onward with the difference of 10 cm whereas short-stemmed varieties are graded from 40 cm – 65 cm with the difference of 5 cm.
2. Leaves should be dark green and healthy. There should not be any yellowing due to micro-nutrient deficiency. Leaves should also be free from dust, chemical residue, chemical injury, spider mites or powdery mildew etc.
3. The flowers should not be bullhead, too open, too tight, with bent neck, damaged by thrips or botrytis etc.
4. Bud size should be representative to the variety
5. The length of the neck should not be too much.

Storage of Cut Flowers

Roses are relatively insensitive to ethylene. Rose cut flowers can be stored dry at 0 to 1^{0}C for 15 days. For dry storage, the flowers are placed in moisture proof boxes and precooled before sealing in the containers. Cut flowers can be stored wet at 2 to 5^{0}C for 5 days. Flowers can be stored at low pressure storage (40-60 mm Hg) for 8 weeks and at modified atmospheric storage (0^{0}C) for 3 to 4 weeks.

Pulsing

Pulsing with 3 per cent sucrose for 18 hours at 20 °C for 45 minutes improves post harvest life and quality of the flowers.

Holding

Vase life of cut roses can be lengthened by keeping in holding solution of 3%

D-fructose + 25 ppm Kinetin or 3% D-fructose + $AgNO_3$ or 3% D-fructose + 300 ppm $NiCl_2$ or 3% sucrose + 200 ppm 8-HQC.

Bud Opening Solution

Tight cut closed buds can be developed to commercial maturity in vase solution containing 1 per cent D-fructose, 100 ppm acetylsalicylic and 250 ppm 8-HQC. Pulsing with 0.2 mM STS for 15 minutes followed by keeping in the holding solution of 300 ppm 8-HQC plus 2 per cent sucrose is effective for bud opening.

Packing and Transport of Flowers

The graded stems are made into bundles of 20 each and tied loosely with rubber band. The buds should be wrapped with 2-ply soft corrugated paper and tied loosely with a rubber band to secure the buds in position. The wrapping paper should protect at least 2-3 cm above the bunch to protect the buds from thorns or from coming in contact with the box. Long stemmed varieties such as Grand Gala or First Red have bigger sized buds. These buds can be packed in five different layers of four flowers in each layer at two different levels. Such staggered bundling is very useful to avoid damage to the buds by mutual pressure. Short stemmed varieties such as Kiss and Sangria have small sized buds which can be packed keeping all the buds in the same plane. The leaves from the lower 4-5 cm portion of the stem are removed and stems put in buckets containing chlorine (50 ppm), aluminium sulphate or citric acid (300 ppm). These chemicals disinfect the stem and check the microbial growth which otherwise could reduce shelf life of the stems. The graded stems are immediately shifted to cold rooms.

Packing Boxes

The bunches are packed in precooled 5-ply teleboxes of fiberboard in such a way that flower heads face in opposite direction i.e. toward the width. The bundles should be placed in rows along the width. The second layer is placed opposite to the first layer.

In the packing box, the layers of flowers are covered with soft paper. Some growers use newspaper to cover the stems but care should be taken to ensure that the newsprint does not leave marks on the stems. After packing the flowers, the box is closed with the lid and strapped properly. Mixing of varieties or stems of varying lengths in the same box should be avoided. The stems should be tightly held inside the boxes so that they are not subjected to jerk movements during transport. The boxes should be properly marked with symbols 'Fresh Flowers' on both sides so that they are handled with care at the airports. Besides, the boxes should have stickers indicating name of the variety, number and length of the stems, it contains.

The dimensions of the packing boxes varies from grower to grower and depends upon the number and length of the stems to be packed and type of the cargo space available in the air craft so that the boxes can fit well into the available space. In most of the commercial units packing box of 100 cm length; 30 cm width; 45 cm height is used to hold 20 number of flowers/ bunch, 16–20 number of bunches/ box with 12-15 kg weight of flower/ box.

Generally, the packing boxes are cooled by placing them open in the cold rooms but the most efficient system is forced air cooling wherein cool air is forced in the boxes through the vents (holes) in the sides of the boxes. When the boxes are to be taken out of the cold room, the vents are closed with self-adhesive BOPP tape to avoid heating of flowers. Vents are, however, kept open if the flowers are to be transported through the refrigerated trucks, to facilitate circulation of cold air through the boxes.

Export

Roses from India are exported to far off places such as Europe and Japan by air transport. The boxes are to be taken to the airport in the refrigerated vans to keep the flowers cool. The International Airport at New Delhi provides cooling facilities to the growers. In case of limited space in the cold room, air conditioning unit of the refrigerated trucks can be plugged to electric points provided by airport authorities to use refrigerated trucks as a make shift cold store at the airport till the flowers are loaded in the aircraft. It should be ensured that the flowers are loaded in a special cabin meant for transport of perishable commodities.

Rehydration of Flowers

After storage during transport, rose stems lose considerable amount of water. Therefore they must be hydrated by putting them in solution of acidifying agents such as aluminium sulphate (300 ppm) or citric acid (300 ppm) for few hours. Before putting the stems in these solutions, basal 2-3 cm stems must be cut underwater to expose new xylem tissues and to remove plugged xylem vessels. Hydration helps the stems to regain turgidity and increases its vase life.

Quality requirement for export

- Pre cooling of flowers
- Free from pest and disease
- Big size with healthy shining leaf
- Long and straight stem (50 – 80 cm)
- Medium to big size flower bud
- Uniform of particular variety and stage of (tight bud or half open) flower in a bunch

The basic requirements of a quality cut rose are

- The straight strong stem capable of holding the flower upright.
- Uniform stem length: Flowers with different stem lengths are not be mixed.
- Size of the flower should be the representative of the cultivar (true to type).
- Uniform stage of development.
- Flower should be free from bruising, injuries, diseases and pests or petal
- discoloration.
- Good, healthy and normal foliage.

Physiological Disorders

1. Bent neck

Insufficient flower stem hardening or insufficient maturation of the stem tissues below the harvested flower results in stem collapse. The reasons are appearance of plugging materials like pectin, cellulose and microbes, extreme temperatures during shipping or storage and water deficiency in the neck tissues. Premature harvesting and excessive water loss also cause bent neck.

2. Bull head or malformed flowers

The central petals of the bud are only partly developed and the bud appears flat. The causes are lack of carbohydrates for petal development, thrips infestation and low temperature during shipping or storage

3. Blind wood

Flowers have sepals and petals, but the reproductive parts are absent or aborted. Blind wood is generally short and thin. But it may attain considerable length and thickness when it develops at the top of the plant. The causes are low temperature and insufficient light during growth and development, insect pests and fungal diseases.

4. Blackening of petals

The major cause is low temperature (20^{0}C at day and 4^0 C at night).

5. Flower bud abortion

No flower bud initiation occurs or if at all it starts, it does not proceed beyond the initiation of pistils and stamen primordials. The causes are competition among growing shoots for assimilates, deficiency of Boron, night temperatures below 15^{0}C and day temperatures above 28^{0}C. Control measures include application of CCC @ 500 ppm and GA @ 100 ppm or application of Boron @ 30 to 60 ppm.

6. Limp neck

The affected flower buds bend down due to the weight of the top. The major causes are insufficient stored energy of flower heads and water stress in the area just below the flower head.

Pests

1. California red scale, *Aonidiella aurantii*

Symptoms

The infested plants bear few small flowers. No rose plant is immune to scales attack.

Management

- Selection of scale free planting material and cutting and burning of infested parts are the operations to reduce the populations
- Spray chlorpyriphos 20 EC @ 2.5 ml/ l or dimethoate 30 EC @ 2.0ml/ l or ethion 50 EC @ 1.0 ml/ l.

- Apply pongamia oil 10% to shoots after pruning to avoid scales infestation
- Apply carbofuran 3G @1.0 kg. a.i./ha (33 kg/ha) after pruning, if the incidence is severe.
- Spray Monocrotophos or Chloropyriphos (1-1.5 ml /lit) + Neem oil + sticker @ 1 ml/lit + 0.5 ml/ lit

2. Thrips (*Scirtothrips dorsalis* Hood *and Rhipiphrothrips cruentatus* Hood)

Symptoms

Both nymphs and adults feed by rasping the tissues and sucking the sap which oozes out from the wounds. Affected leaves are deformed with brown or silvery patches or burnt margins, The affected leaves get distorted, wither and drop down. In case of severe infestation, 100eggs. The nymphs hatch in 2-7 days and starts feeding on the plant parts, fully grown nymphs pupate in soil and emerge as adults in 2-5 days depending on temperature.

Management

- Spray acephate 75 SP @ g/ 1 or dimethoate 30 EC @ 2.0 ml/ 1 followed with 1% pongamia oil, 2-3times at fortnightly interval with onset of new flush.
- Drench the soil with chlorpyriphos 20 EC @ 5.0 ml/ 1 at fortnightly interval to kill pupae in soil
- Under polyhouse cultivation, spray with fipronil 5 SC @ 1.5 ml/l or imidacloprid 200 SL @ 0.4 ml/I or acephate 74 SP @ 1.5 @ 1.5 g/ 1
- Spray Confidor or Agas or Bullet or Polo (1-1.5 ml or g/ lit) + Sandavit @ 0.3 ml/lit

3. Aphid, *Macrosiphum rosae* Linnaeus

Symptoms

Both adults and nymphs found in clusters on the tender portions of shoots, buds, flowers and leaves. They feed by sucking the cell sap. As a result, tender shoots wither, buds fall prematurely and the flowers show malformation and fading. Severe damage to the top of the plant may reduce the number of flowers produced. Aphids also excrete honey dew on which sooty mold grows. Aphids feeding on flowers make them unmarketable and their presence is a nuisance on the plants because they leave cast skins stuck to the plant when they molt, which affect the value of the plant/flower.

Management

- Spray 1% neem or pongamia oil or dimethoate 30 EC @ 2.0 ml/ l
- If the incidence is severe, spray imidacloprid 200 SL @ 0.4ml/ l or cartap hydrochloride 50 SP @ 1.0 g/ l
- Spray of *Verticillium lecanii* at 3.0 g/ l during evening hours is also effective against aphids

4. Bud borer, *Helicoverpa armigera* (Hubner)

Symptoms

Hatched larvae bore into buds by making holes and feed on petals. Grown up stages damage the flowers.

Management

- Collection and destruction of mature larvae reduce the population build up
- Spray methyl parathion 50 EC @ 1.0 ml/ l or fenvalerate 20 EC @ 0.5 ml/l in combination with difulubenzuron 25 WP @ 2.0 g / l after appearance of eggs on tender buds.
- Alternatively spray neem seed kernel extract (NSKE) 4% at weekly intervals
- Spraying of HaNPV @ 250 LE/ha

5. Whitefly, *Bemisia tabaci (Hemiptera: Aleyrodidae*)

Symptoms

It affects the growth of the plants resulting in production of small flowers

Management

- Removal and burnings of heavily infested leaves checks pest build up
- Clean cultivation and use of insect proof nets helps in prevention of whitefly incidence
- Install yellow sticky traps to monitor adult flies activities
- Spray with acephate 75 SP @ 1.5 g/ l or fortnightly interval alternating with pongamia oil @ 10.0 ml/ l
- Spray of *Beauveria bassiana* or *Verticillium lecanii* formulation @ 2.0 ml / l. If the activity of adults is more, spray with dichlorvos 76 EC @ 5.0 ml/ l followed by deltamethrin 2.8 EC @ 1.0 ml/l at 5-7 days interval

- Regular water spray on the plants and yellow trap
- Spray harrier or polo or lannate @ 1-1.5 ml or g /lit + Neem oil + sticker @ 1 ml/lit + 0.5 ml/ lit spray

6. Tobacco caterpillar, *Spodoptera litura*

Symptoms

Brown coloured mature larvae damage growing buds and flowers during nights resulting in qualitative loss to flowers.

Management

- Collect and destroy egg masses and gregarious early instar larvae
- Spray quinalphos 25 EC @ 2.0 ml/I at fortnightly interval.
- Spray indoxacarb 14.5 SC @ 1.0 ml/ l or thiodicarb 75 WP @ 1.0 g / l if the incidence is severe
- Spraying of SINPV @ 250 LE/ha followed by neem formulations 1.0 – 2.0 ml / l are also effective
- Spread poison bait made of wheat or rice bran, jaggary and chlorpyriphos (10:1:0.5) for killing grown up larvae
- Methyl parathion + sticker @ 1ml/lit + 0.5 ml/lit spray
- Quinalphos + Sticker @ 2ml/lit + 0.5 ml/lit spray

7. Spider mite (*Tetranychus cinnabarinus)*

Symptoms

Symptoms are white specks on leaves which later coalesce and produce white patches resulting in reduced photosynthetic activity.

Management

- Cutting and burning of severely infested plant parts, proper irrigation and clean cultivation reduce incidence of mites.
- Thorough spray of jet of water to dislodge mites from their webs and plants followed by application of dicofol 18.5 EC @ 2.5 ml/ l or wettable sulphur 80 WP @ 3 g / l or profenofos 50 EC @ 1 ml/ l or ethion 50 EC @ 1.0 ml/I followed by pongamia or neem oil at 5 ml / l
- Spraying Verticillium lecanii at 5.0 g/ l during evening hours.
- Spray abamectin 1.9 EC @ 0.5 ml/ l at followed by flufenoxuron 10 DC @ 1.0ml/I or fenazaquin 10 EC @ 1.0 ml/ l or diafenthiuron 50 SC @ 0.6 ml/I, if necessary on the crop meant for export purpose.

- Regular water spray on the plants
- Maiden or Mite block or Dacomain (1-1.5 ml/lit) + Neem oil + sticker @ 1 ml/lit + 0.5 ml/ lit spray

Diseases

1. Black leaf spot (*Diplocarpon rosae*)

Symptoms

Symptoms appear in the form of circular black spots ranging from 1/16 inch to ½ inch in diameter on the surface of leaves. The spots are frequently surrounded by a yellow hallow. Infected leaves characteristically turn yellow and fall prematurely.

This leaf spot can be distinguished from other leaf spots by characteristic fringed margin. The disease can cause almost complete defoliation of bushes. Cane may also get infected and develop reddish purple raised irregular spots. The infected bush becomes weak and prone to attack by cane die back, stem canker and winter injury.

Management

- Observe good hygiene in the field.
- Collect diseased plant debris with a garden rake and destroy it.
- Give pruning cuts at least one inch deep into the healthy wood.
- Spray the bushes with fungicides, such a carbendazim (0.1%), chlorothalonil (0.2%), mancozeb (0.2%), kresoxim-methyl (0.1%), or tebuconazole (0.1%). Applications should begin as soon as the new leaves appear or at first appearance of black spots. Sprays may be repeated at an interval of 10 days or more often depending upon the severity of infection.
- Improve ventilation and avoid watering in the late afternoon.
- Propiconazole + Sticker @ 1 ml/lit + 0.3 ml/lit spray.
- Ridomil or Aliette or Blitox + Sticker @ 1-1.5 g /lit + 0.3ml/lit spray.

2. Powdery Mildew, (*Podosphaera pannosa*)

Symptoms

Leaves, buds and stems get covered with a white powdery coating, The infection can cause young leaves to curl and turn purple. Young canes may be distorted and dwarfed. If seriously infected, they can die. Severely infected buds fail to open.

Management

- Spray plants with dinocap (Karathene (0.05%), penconazole (0.05%), propiconazole (0.05%) or fenarimol (0.1%) at fortnightly intervals
- Use disease resistant varieties
- Remove the affected plant parts (leaves, stems and buds)
- Regular spray of *Bacillus sp,*
- Nativo or Melody Duo or Folicur (1-1.5 ml or g/lit) + sticker @ 0.5ml/lit spray

3. Stem canker/ Dieback (*Diplodia rosarum*)

Symptoms

As the name implies, the disease causes death of plants from downwards. Infection normally spreads from the pruned end of the twigs and extends only a few centimeters below the cut end. During severe infection, however, the disease may spread right down to the base of the canes and subsequently the whole plant resulting in its death.

Management

- Observe sanitation in and around the poly house.
- Prune and destroy dead and diseased plant parts periodically.
- Disinfect secateurs with 70 per cent alcohol before giving pruning cut to each cane. Mercuric chloride has to be avoided for disinfection, as the roses are particularly sensitive to this compound
- Apply Bordeaux paste (copper sulphate 1 kg, unslaked lime 1.5kg, water 10-15 liters) or Chaubattia paint (copper carbonate 4 parts, red lead 4 parts, raw linseed oil 5 parts) to cut ends at time of pruning of bushes.
- The crop may be sprayed with copperoxychloride (0.3%) or mancozeb (0.25%) also at the time of pruning of bushes.
- Prune the infected stem below 5 cm of healthy stem. Bordeaux mixture pasting the pruned end
- Spray Bavistin or Blitox or Aliette or Acrobat (1-1.5 g /lit) + sticker @ 0.5ml/lit

4. Blossom blight, *Botrytis cinerea*

Symptoms

Infected buds usually fail to open and become covered with grey to grayish brown growth of the fungus. Infection of open flowers also occurs. Small, more or less circular, lesions form on petals that may develop into blotches and rot the flowers. The disease causes flower buds to droop and remain closed. Buds turn brown and decay. Sometimes partially opened buds are attacked and an entire flower may be covered by gray fungus.

Management

- Practice crop sanitation
- Destroy dead and diseased plant parts
- Avoid moisture condensation in the polyhouse
- Adjust temperature to 21°C and relative humidity below 85 % in the green house area
- Spray plants with mancozeb (0.2%) or chlorothlanil (0.2%) at suitable intervals. Sprays of the fungicides vinclozolin (0.2%) are also very effective.

5. Downy mildew (*Peronospora sparsa)*

Symptoms

Disease symptoms appear in the form of purplish-red to dark brown irregular lesions, delimited by veins, on the under surface of leaves. Stems, petioles and flower stalks may also develop purple marks.

Management

- Hygiene in and around the poly house
- Periodic removal of diseased leaves/infected parts.
- Alternate sprays with mancozeb (0.25%), metalazyl-mancozeb (0.25%), fosetyl-aluminium (0.3%) or ziram (0.3%) at 10-14 day intervals
- Decrease the relative humidity in the green house is the more effective
- Ridomil + Sticker @ 1.5 gm/lit + 0.3 ml/lit spray
- Bavistin or Blitox (1-1.5 g /lit) + sticker @ 0.5ml/lit spary

6. Crown gall (*Agrobacterium tumefaciens)*

Symptoms

The disease is characterized by large lumps at the base of the plant stem or on roots. Galls may appear higher on stems as the disease progresses. Galls are soft compared to surrounding plant tissues. If the disease affects the plant while it is young, the plant may be affected to the degree where it will not produce blooms. All affected plants wilt readily and grow poorly.

Management

- Avoid injury to the plants
- Purchase disease-free stock plants from a reputed supplier
- Rogue out diseased plants early.
- Cultural control through prevention of introduction/elimination is the best method
- Chemical control of the disease is not available
- Biological control of the disease with A. *radiobacter* strain K84 or K1026 reported successful.
- Remove and destroy the plant, clean nursery plant, suitable rootstocks

7. Mosaic, *rose mosaic virus, rose ring pattern virus*

Symptoms

This disease is characterized by prominent area on the leaves. The patterns vary considerably, ranging between all-over fine blotches to patterns of lines in waves. These symptoms may not appear on all leaves.

Management

- Remove and destroy diseased plants
- Propagate cutting from healthy plants only
- Disinfect cutting tools between stock plants from which cuttings are being taken
- Use bud woods from plants which received heat treatment (33° C, 4 wks)
- Meristem culture can also help to eliminate virus.

Zeitfracht Medien GmbH
Ferdinand-Jühlke-Straße 7
99095 Erfurt, Deutschland
produktsicherheit@kolibri360.de